# ESTABLISHING A TRAINING PROGRAMME ON THE SAFETY REVIEW OF NUCLEAR INSTALLATION SITES

The following States are Members of the International Atomic Energy Agency:

AFGHANISTAN
ALBANIA
ALGERIA
ANGOLA
ANTIGUA AND BARBUDA
ARGENTINA
ARMENIA
AUSTRALIA
AUSTRIA
AZERBAIJAN
BAHAMAS
BAHRAIN
BANGLADESH
BARBADOS
BELARUS
BELGIUM
BELIZE
BENIN
BOLIVIA, PLURINATIONAL STATE OF
BOSNIA AND HERZEGOVINA
BOTSWANA
BRAZIL
BRUNEI DARUSSALAM
BULGARIA
BURKINA FASO
BURUNDI
CABO VERDE
CAMBODIA
CAMEROON
CANADA
CENTRAL AFRICAN REPUBLIC
CHAD
CHILE
CHINA
COLOMBIA
COMOROS
CONGO
COOK ISLANDS
COSTA RICA
CÔTE D'IVOIRE
CROATIA
CUBA
CYPRUS
CZECH REPUBLIC
DEMOCRATIC REPUBLIC OF THE CONGO
DENMARK
DJIBOUTI
DOMINICA
DOMINICAN REPUBLIC
ECUADOR
EGYPT
EL SALVADOR
ERITREA
ESTONIA
ESWATINI
ETHIOPIA
FIJI
FINLAND
FRANCE
GABON
GAMBIA, THE
GEORGIA
GERMANY
GHANA
GREECE
GRENADA
GUATEMALA
GUINEA
GUYANA
HAITI
HOLY SEE
HONDURAS
HUNGARY
ICELAND
INDIA
INDONESIA
IRAN, ISLAMIC REPUBLIC OF
IRAQ
IRELAND
ISRAEL
ITALY
JAMAICA
JAPAN
JORDAN
KAZAKHSTAN
KENYA
KOREA, REPUBLIC OF
KUWAIT
KYRGYZSTAN
LAO PEOPLE'S DEMOCRATIC REPUBLIC
LATVIA
LEBANON
LESOTHO
LIBERIA
LIBYA
LIECHTENSTEIN
LITHUANIA
LUXEMBOURG
MADAGASCAR
MALAWI
MALAYSIA
MALI
MALTA
MARSHALL ISLANDS
MAURITANIA
MAURITIUS
MEXICO
MONACO
MONGOLIA
MONTENEGRO
MOROCCO
MOZAMBIQUE
MYANMAR
NAMIBIA
NEPAL
NETHERLANDS, KINGDOM OF THE
NEW ZEALAND
NICARAGUA
NIGER
NIGERIA
NORTH MACEDONIA
NORWAY
OMAN
PAKISTAN
PALAU
PANAMA
PAPUA NEW GUINEA
PARAGUAY
PERU
PHILIPPINES
POLAND
PORTUGAL
QATAR
REPUBLIC OF MOLDOVA
ROMANIA
RUSSIAN FEDERATION
RWANDA
SAINT KITTS AND NEVIS
SAINT LUCIA
SAINT VINCENT AND THE GRENADINES
SAMOA
SAN MARINO
SAUDI ARABIA
SENEGAL
SERBIA
SEYCHELLES
SIERRA LEONE
SINGAPORE
SLOVAKIA
SLOVENIA
SOMALIA
SOUTH AFRICA
SPAIN
SRI LANKA
SUDAN
SWEDEN
SWITZERLAND
SYRIAN ARAB REPUBLIC
TAJIKISTAN
THAILAND
TOGO
TONGA
TRINIDAD AND TOBAGO
TUNISIA
TÜRKİYE
TURKMENISTAN
UGANDA
UKRAINE
UNITED ARAB EMIRATES
UNITED KINGDOM OF GREAT BRITAIN AND NORTHERN IRELAND
UNITED REPUBLIC OF TANZANIA
UNITED STATES OF AMERICA
URUGUAY
UZBEKISTAN
VANUATU
VENEZUELA, BOLIVARIAN REPUBLIC OF
VIET NAM
YEMEN
ZAMBIA
ZIMBABWE

The Agency's Statute was approved on 23 October 1956 by the Conference on the Statute of the IAEA held at United Nations Headquarters, New York; it entered into force on 29 July 1957. The Headquarters of the Agency are situated in Vienna. Its principal objective is "to accelerate and enlarge the contribution of atomic energy to peace, health and prosperity throughout the world".

IAEA-TECDOC-2083

# ESTABLISHING A TRAINING PROGRAMME ON THE SAFETY REVIEW OF NUCLEAR INSTALLATION SITES

## A HANDBOOK FOR REGULATORY BODIES IN EMBARKING COUNTRIES

INTERNATIONAL ATOMIC ENERGY AGENCY
VIENNA, 2025

For further information on this publication, please contact:

External Events Safety Section
International Atomic Energy Agency
Vienna International Centre
PO Box 100
1400 Vienna, Austria
Email: Official.Mail@iaea.org





Printed by the IAEA in Austria
April 2025
https://doi.org/10.61092/iaea.pqbx-stq5

**IAEA Library Cataloguing in Publication Data**

Names: International Atomic Energy Agency.
Title: Establishing a training programme on the safety review of nuclear installation sites / International Atomic Energy Agency.
Description: Vienna : International Atomic Energy Agency, 2025. | Series: IAEA TECDOC series, ISSN 1011-4289 ; no. 2083 | Includes bibliographical references.
Identifiers: IAEAL 25-01746 | ISBN 978-92-0-105325-1 (paperback : alk. paper) | ISBN 978-92-0-105225-4 (pdf)
Subjects: LCSH: Nuclear facilities — Employees — Training of. | Nuclear facilities — Safety measures. | Nuclear facilities — International cooperation. | Nuclear power plant operators — Training of.

# FOREWORD

In 2020, the IAEA launched an extrabudgetary project entitled Capacity Building on Site Safety Review and Assessment in Embarking Countries, aimed at enhancing the capacity and competence of the managers and technical personnel of regulatory bodies who will review the safety of nuclear installation sites against external events in Member States embarking on new nuclear programmes.

The project led to the production of a self-assessment module for site survey, selection and evaluation; a framework for a training workshop; and a safety analysis report review practice module that consists of review guidance, sample safety analysis reports and safety evaluation reports on site characterization, and templates and procedures for the practice. As part of the project, seven pilot Member States were selected in which these products were tested and optimized through national and interregional training workshops. The final products from the project have been compiled in IAEA publications or working documents for training.

This publication, developed under the Generic RoadMap project on developing infrastructure for nuclear safety of the first reactor, incorporates insights gathered from the capacity building project and supplements them with case studies drawn from education and training programmes in operation across four selected Member States: the Republic of Korea, Türkiye, the United Arab Emirates and the United States of America. By integrating these diverse experiences, the publication serves as a comprehensive information source and practical tool for enhancing the capacity of regulatory bodies in reviewing the safety of nuclear installation sites in embarking countries.

This publication is intended to provide the regulatory bodies of embarking countries with information on current practices and experiences to support implementation of their capacity building programmes for the review of nuclear installation site safety. It addresses capacity building fundamentals and training programme development and provides case studies of the capacity building programmes in the IAEA and several Member States with advanced programmes, focusing on the review of nuclear installation site safety against external events. The publication is expected to be of benefit to regulatory bodies and operating organizations establishing or building their human capacity to conduct activities related to the review of nuclear installation site safety.

The IAEA is grateful to the European Commission and the United States of America for their financial support for this publication, and to all the experts who contributed to its drafting and review, in particular M. Dubinsky (Israel) and S. Lee (Republic of Korea). The IAEA officer responsible for this publication was H. Lee of the Division of Nuclear Installation Safety.

CONTENTS

# 1. INTRODUCTION

## 1.1. BACKGROUND

The Statute of the IAEA includes the establishment, as well as provision for the application, of safety standards for the protection of health and minimization of danger to life and property. Education and training play a key role in facilitating the application of safety standards in IAEA Member States and in globally strengthening nuclear and radiation safety. The education and training activities of the IAEA follow the resolutions of the General Conference and reflect IAEA safety standards and related publications.

In the wake of the accident at the Fukushima Daiichi nuclear power plant (NPP) in 2011, the IAEA developed its Action Plan on Nuclear Safety [1], which was approved by the Board of Governors and unanimously endorsed by the General Conference at its 55th session. Capacity building in Member States is one of the twelve main actions identified in the Action Plan. The IAEA devised an integrated capacity building concept that addresses four main pillars: education and training, human resource development, knowledge management and knowledge networks.

In 2012, the IAEA conducted a review of achievements in education and training on nuclear safety over the period 2001–2012 and developed a strategic approach to education and training in nuclear safety for the period 2013–2020 [2]. The strategic approach includes a summary of the key achievements in education and training on nuclear safety over the period 2001–2012 and is in line with and supports the IAEA Action Plan on Nuclear Safety [1]. This strategic approach highlights the importance of Member States taking responsibility for nuclear safety, by developing and implementing national strategies to that end, and warns that a high level of national commitment will be needed to put in place the mechanisms for building national capacity in nuclear safety.

The basic components of this strategic approach are (a) national strategies for capacity building, (b) capacity building mechanisms, (c) effective use of networking, regional cooperation and regional training centres, (d) management systems, competence and knowledge management, and (e) monitoring process.

In 2015, the IAEA published a report on capacity building [3] to highlight the relevant lessons learned in the light of the Fukushima Daiichi accident for strengthening capacity building for nuclear safety in Member States. The report emphasized that capacity building is a major first step in the process of ensuring a sustainable supply of competent human resources capable of applying nuclear technologies in a safe, responsible and sustainable manner. As stated in IAEA Safety Standards Series No. SF-1, Fundamental Safety Principles [4], the licensee, which retains the prime responsibility for safety throughout the lifetime of facilities and activities, is responsible for establishing and maintaining the necessary competences.

In 2020, the IAEA's External Events Safety Section (EESS) launched the extrabudgetary project entitled 'Capacity building on site safety review and assessment in embarking countries' (hereinafter referred to as the CB Project). The CB Project is aimed at enhancing the capacity and competence of embarking countries in the evaluation and review of site safety for new nuclear installations by development of an e-learning platform for general disciplinary training; self-assessment tools and capacity building action plans for country-specific training; on-the-job training tools, including capacity building training manuals and sample training documents; and a manual for the review of the site evaluation chapter of the safety analysis report (SAR).

In 2022, under the Generic Roadmap project on developing infrastructure for nuclear safety of the first reactor, the IAEA launched an initiative to produce a publication on education and training programmes for regulatory bodies in embarking countries, focusing on the review of safety of nuclear installation sites.

## 1.2. OBJECTIVE

The objective of this publication is to provide a practical reference, based on current practices and experiences, for regulatory bodies of embarking countries to support implementation of their capacity building programmes on the review of nuclear installation site safety. Other organizations may also find value in consulting this publication.

This publication forms part of a tool kit to help managers and staff become more familiar with the competencies needed to review nuclear installation site safety. It is also designed to help managers and staff understand how competencies can be used to support more effective recruitment and selection strategies. Over time, selecting candidates on the basis of these competencies and further developing these competencies in staff members will also help foster superior performance.

## 1.3. SCOPE

This publication is in line with the requirements established in IAEA Safety Standards Series No. SSR-1, Site Evaluation for Nuclear Installations [5], and with the requirements, recommendations and information on regulatory confidence and capacity building contained in Refs [6–15].

The publication applies to capacity building activities, with particular focus on the education and training area, for the managers and technical personnel of regulatory bodies in Member States embarking on new nuclear programmes, who will evaluate and review the safety of nuclear installation sites against external events. The publication addresses capacity building fundamentals and training programme development, and includes case studies of the capacity building programmes in the Agency and several Member States with advanced programmes.

The direct beneficiaries are regulatory bodies and operating organizations that are establishing or building their human capacity to conduct activities to review site safety for a nuclear installation.

## 1.4. STRUCTURE

This publication consists of six sections in the main body, one appendix and four annexes.

Section 1 describes the background, objective, scope and structure of this publication.

Section 2 explains the fundamentals of developing a national strategy for education and training and the key concepts of capacity building programme design.

Section 3 focuses on considerations in developing an education and training programme, taking into account safety standards on nuclear installation site safety. It addresses the fundamental safety principles and the legal and regulatory framework, as well as licensing, inspection and enforcement in relation to the site safety of nuclear installations.

Section 4 describes the IAEA's training programme to build capacity in the review of nuclear installation site safety, including self-assessment, establishment of a capacity building action plan, design and implementation of a training programme, evaluation and feedback, and post-training activities.

Section 5 presents good practices in selected Member States relating to human resources, capacity building training programmes and knowledge management for nuclear installation site safety.

Section 6 provides a summary of the publication.

The Appendix contains a list of suggested areas to be included in a capacity building training programme on the review of nuclear installation site safety.

Annexes I and II provide supplementary details of the capacity building programmes offered by the IAEA and selected Member States. Annex III provides a sample inspection procedure and Annex IV provides a sample train-the-trainer programme.

## 2. CAPACITY BUILDING FUNDAMENTALS

A competence is generally considered as a combination of knowledge, skills and attitudes (KSAs) [13]. The term 'capacity building' refers to mechanisms designed to improve the capacity of individuals or organizations to perform a given task.

The competence of the regulatory body to perform its responsibilities are addressed in para. 4.13 of IAEA Safety Standards Series No. GSR Part 1 (Rev. 1), Governmental, Legal and Regulatory Framework for Safety [6], as follows:

> "A process shall be established to develop and maintain the necessary competence and skills of staff of the regulatory body, as an element of knowledge management. This process shall include the development of a specific training programme on the basis of an analysis of the necessary competence and skills. The training programme shall cover principles, concepts and technological aspects, as well as the procedures followed by the regulatory body for assessing applications for authorization, for inspecting facilities and activities, and for enforcing regulatory requirements."

In accordance with Requirement 19 of GSR Part 1 (Rev. 1) [6], the regulatory body is required to establish a management system for the management of its regulatory activities. Competence management needs to be integrated into this management system, including such factors as the systematic approach to training, quality assurance, continuous improvement and staff development, as discussed in IAEA Safety Standards Series No. GSG-12, Organization, Management and Staffing of the Regulatory Body for Safety [10].

In order to assist the regulatory body in developing a high level of competence, the IAEA has published several guidance and informational publications [8–20] on capacity building for nuclear safety, including on topics such as human resource development and capacity building management.

### 2.1. NATIONAL STRATEGY FOR EDUCATION AND TRAINING

Section 5.2 of Ref. [3] states that:

> "Education and training infrastructure and processes are fundamental to the capacity building strategy of Member States, as they provide a structure to develop the capacity of those individuals involved in the application of nuclear technologies. Education and training provide the basis for human resource development, knowledge management and knowledge networking. A systematic approach to training, including quality assurance and continuous improvement, needs to be a component of the management system of all organizations relevant for nuclear safety."

Section C.2.1 of Ref. [2] states that:

> "A national strategy for capacity building needs to be developed, implemented and periodically assessed… Member States with nuclear power programmes and those planning to embark on a nuclear power programme should strengthen, develop, maintain and implement their capacity building programmes."

The four main pillars of capacity building — education and training, human resource development, knowledge management and knowledge networks — are discussed in Ref. [2]. In the design and implementation of an education and training programme, it is necessary to assess and prioritize education and training needs for nuclear facilities and activities, the available national resources and the support of regional or international resources. Following implementation, the education and training programme may be evaluated to identify further education and training needs. The systematic approach shown in Fig. 1 for an education and training programme is recommended in Ref. [2].

*FIG. 1. Phases to establish and maintain a national strategy for education and training (adapted from Ref. [2]).*

To establish a strong reservoir of personnel with the necessary skills, training in the sciences, technology, engineering and mathematics (the so-called 'STEM' programme) needs to be established in the primary and secondary schools, so that students are ready for university education in the areas of study necessary for nuclear power programme development. This is an investment that the Member State may need to make early on. Beyond internal knowledge development, it may be beneficial to transfer knowledge from the education and training activities of external sources to the regulatory body personnel in Member States embarking on a nuclear power programme.

For continuous improvement in safety-related knowledge, the development of a knowledge network supports education, training, and human resource development, while also providing access to lessons learned from the capacity building experience of others.

## 2.2. CAPACITY BUILDING PROGRAMME

Capacity building programmes for nuclear safety, and specifically education and training programmes, might be based on a systematic approach that includes provisions for planning, assessment, implementation, documentation, and evaluation and feedback. This subsection describes a possible methodology for developing an effective capacity building programme.

### 2.2.1. Competence management

The organizational structure, management and staffing of the regulatory body are important factors for ensuring the control of facilities and activities. Changes in staffing, including recruitment, retirement and other departures, also need to be considered in competence management.

Paragraph 6.16 of IAEA Safety Standards Series GSG-12, Organization, Management and Staffing of the Regulatory Body for Safety [10] states that:

"6.16. The competence management process may include the following subprocesses [13]:

— Analysis of competence needs:
  - Task analysis leading to determination of the necessary competences;
  - Analysis of existing competences within the regulatory body;
  - Gap analysis (personal performance review and assessment).
— Prioritization of competence needs and filling competence gaps:
  - Recruitment and human resources planning;
  - Staff training and development;
  - Management of external expert support.
— Knowledge capture and management.
— Reviews and audits of competence management and feedback."

Reference [20] presents the concepts, drivers and benefits of managing nuclear safety knowledge, with a focus on specific considerations in Member States, taking into account their current national approaches and experience. Reference [20] also provides practical guidance to support Member States in implementing IAEA safety standards relating to managing nuclear safety knowledge at the national level.

Reference [21] describes the Canadian Nuclear Safety Commission method for developing a radiation safety training programme. This method could be valid for capacity building for a nuclear installation site safety programme. Reference [21] presents the six main steps for developing a radiation safety training programme that identifies and responds to activity-specific training needs, based on the systematic approach to training, as follows:

(1) Identification of specific training needs;
(2) Training programme design;
(3) Development of training materials;
(4) Instruction (including trainee assessment);
(5) Training programme evaluation;
(6) Continuous improvement and continued training.

### 2.2.2. Competence model

A competence model provides a structure for assessing the competence needs of an organization. The regulatory body needs to establish the necessary competences for its regulatory activities, determine the necessary levels of competences and establish standards for evaluation of competences as part of the capacity building programme. Reference [13] provides a general competence model for this purpose, using a quadrant structure. This quadrant model is reproduced in Fig. 2.

Section 3.1. of Ref. [13] states that:

> "The competence model is based on a quadrant structure: Quadrant 1 contains the competences related to the legal, regulatory and organizational basis; Quadrant 2, the competences related to technical disciplines; Quadrant 3, the competences related to a regulatory body's practices; and Quadrant 4, personal and behavioural competences… Each quadrant comprises a set of competence areas with a set of specific competences (KSAs)."

While the quadrant model is widely applicable across regulatory bodies and organizations, it is important to note that the competency areas have been predominantly designed with regulatory bodies in mind.

Therefore, adjustments may be necessary to better align them with the specific needs of other types of organisations.

Examples of training programmes as per the quadrant model of competences for regulatory bodies are provided in Annex I and Annex II. Annex I describes the training programmes developed and conducted by the IAEA, whereas Annex II provides several examples of training programmes from Member States

| **1. Competences related to the legal, regulatory and organizational basis**<br>1.1. Legal basis<br>1.2. Regulatory policies and approaches<br>1.3. Regulations and regulatory guides<br>1.4. Management system | **2. Technical disciplines competences**<br>2.1. Basic science and technology<br>2.2. Applied science and technology<br>2.3. Specialized science and technology |
|---|---|
| **3. Competences related to a regulatory body's practices**<br>3.1. Review and assessment<br>3.2. Authorization<br>3.3. Inspection<br>3.4. Enforcement<br>3.5. Development of regulations and guides | **4. Personal and behavioural competences**<br>4.1. Analytical thinking and problem solving<br>4.2. Personal effectiveness andself-management<br>4.3. Communication<br>4.4. Teamwork<br>4.5. Managerial and leadership competences<br>4.6. Safety culture |

*FIG. 2. Quadrant model of competences for regulatory bodies (reproduced from Ref. [13]).*

### 2.2.3. Planning of competences and staffing

In order to plan competences and staffing, the regulatory body might consider and assess the regulatory functions that need to be performed. A strategic plan may need to be developed for developing and maintaining competences, including training and development, staffing plans, and the use of external expert support.

IAEA Safety Standards Series No. SSG-75, Recruitment, Qualification and Training of Personnel for Nuclear Power Plants [12], provides recommendations on systematically recruiting, qualifying, and training personnel for nuclear power plants to ensure safe and efficient operations. Additionally, Ref. [13] offers guidance on planning and managing the competences of nuclear regulators. Together, these references outline a strategic approach to human resource development within organizations involved in nuclear safety, including a structured programme for competence acquisition and development.

### 2.2.4. Systematic competence analysis

The IAEA developed the Systematic Assessment of Regulatory Competence Needs (SARCoN) methodology [16] and software tool, which provides information on how to gather and analyse information on competence needs, identify existing competences, and implement gap analyses. SARCoN includes a comprehensive set of questions to identify gaps in KSAs in each of the four competence areas of the quadrant model shown in Fig. 2.

The competences needed by the regulatory body to fulfil its regulatory functions are identified by means of a systematic analysis of competence needs based on the regulatory body's function and processes. A gap analysis is used to compare existing competences with needs, and the results are used to develop priorities for action, as shown in Fig. 3.

The competence analysis process presented in Ref. [13] can be applied in light of the objective and scope of this publication as follows:

(a) Identifying the functions and related tasks of the regulatory body

In this publication, the assumption is made that the main functions of the nuclear regulatory body are to be responsible for oversight of nuclear safety and public health (and environmental) safety, and to review the SAR and the radiological environmental impact assessment (EIA). Thus, in line with GSG-12 [10], the core regulatory functions are development of regulations and guides; notification and authorization; review and assessment of facilities and activities; inspection of facilities and activities; enforcement; emergency preparedness and response; and communication and consultation with interested parties.

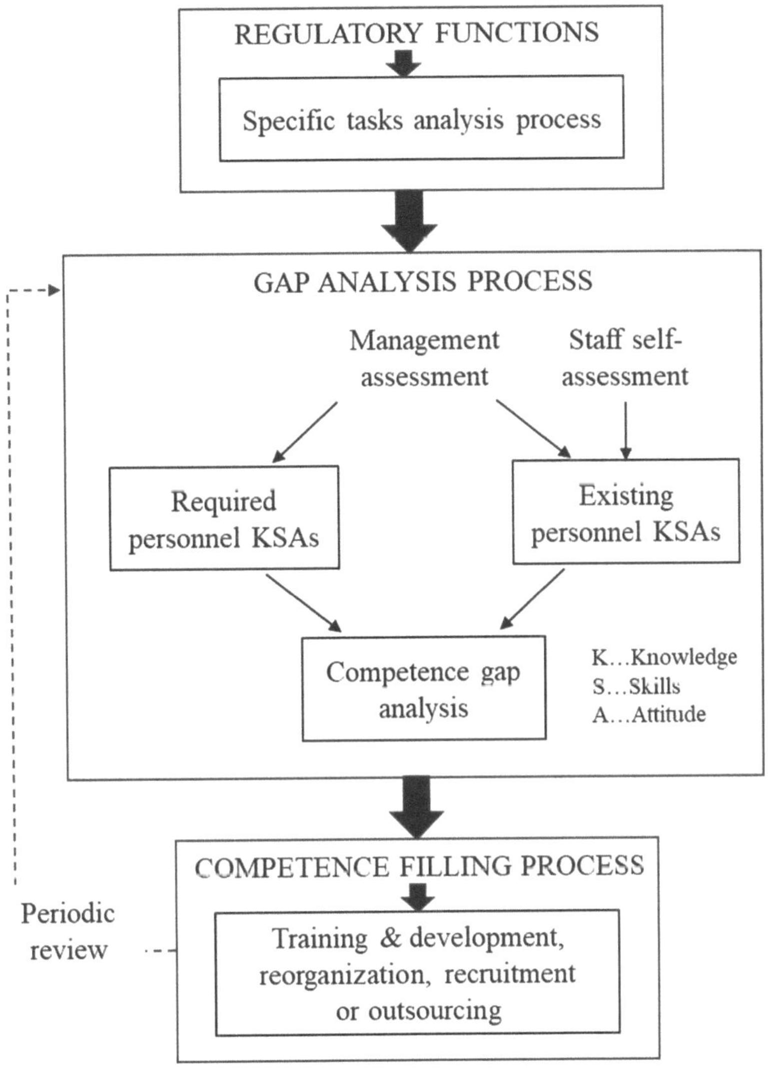


*FIG. 3. Flow chart for competence analysis (reproduced from Ref. [13]).*

(b) Identifying the necessary competences associated with tasks

Tasks involve specific skills. The department responsible for developing a capacity building programme has the responsibility to identify the skills needed, and to determine the necessary level of skill, considering the organization's structure and the tasks that need to be performed. Future needs and aspirations can also be taken into account. While the model suggested in Ref. [13] rates competence as high, medium or basic, regulatory bodies may choose to use more than three levels and might select different definitions. For example, the self-assessment module developed by the IAEA for identifying the competence of regulatory bodies and interested organizations in embarking countries in the review of nuclear installation site safety uses five levels: 1 (very weak), 2 (weak), 3 (average), 4 (strong) and 5 (very strong). The topical areas rated 1 to 3 are considered as potential areas for training or other types of capacity building.

(c) Analysing competence gaps

As a next step, a self-assessment can be performed by an individual or team. The existing competence is assessed through personal performance reviews, examination of each individual's competence, and discussion of proposed competence development. The gaps in competence can be analysed overall at any organizational level and for the organization as a whole.

(d) Prioritizing competence gaps

Not all competence gaps are equally critical or urgent. Some may have a more significant impact on regulatory compliance, operational efficiency, or overall performance. Prioritization involves ranking these gaps based on their importance, potential consequences and urgency. Once the competence gaps are identified for the individual, team or organization, they need to be prioritized to plan for capacity building, taking into account their importance to the regulatory functions. Typical ways of filling competence gaps might be personal development plans, reallocation of duties, recruitment, training, or use of external support.

### 2.2.5. Evaluation and monitoring of the capacity building programme

The capacity building programme needs to be flexible and adjustable to allow implementation of new methods or processes to achieve ongoing improvement within the regulatory body's management system.

The capacity building programme needs to be continually evaluated on the basis of results collected during previous phases. The evaluation provides feedback that can facilitate improvement in the capacity building programme. Sources of feedback include course evaluation, self-evaluation of performance improvement, feedback from managers, and periodic gap analysis activities. An appropriate monitoring programme will need to be established including performance indicators to give an overall assessment of capacity building programme implementation and effectiveness.

## 3. DEVELOPING A TRAINING PROGRAMME

Training is defined, developed and maintained through an iterative and interactive series of steps, leading from the identification of a training need to confirmation that the need has been met. The training programme presented in this publication is designed to be performance oriented. It focuses on imparting essential knowledge, as well as developing the skills and safety-related attributes needed to perform tasks and to meet safety-related needs throughout the life cycle of the nuclear installation.

This section first addresses general training for nuclear safety, followed by specific training for the review of nuclear installation site safety. The general training subsections are necessary for the understanding of fundamental nuclear safety and regulatory activities to ensure the safety of nuclear facilities and activities. They include such topics as the fundamental safety principles, the legal and regulatory framework, the licensing process, review and assessment, inspection, and enforcement. These training topics are primarily based on Ref. [4]). Specific training topics for the review of nuclear installation site safety are mostly based on Refs. [23, 30].

Following are key areas of the training programme for general nuclear safety as well as the specific training for the site safety of nuclear installations. The Appendix provides a detailed list of possible training topics for site evaluation. Additionally, several examples of training programmes including the list of potential topics in the training manuals are provided in Annex I.

## 3.1. FUNDAMENTAL SAFETY PRINCIPLES

The Fundamental Safety Principles established in SF-1 [4] constitute the basis to establish safety requirements for protection against exposure to ionizing radiation under the IAEA's safety standards programme and provide the rationale for its wider safety related programme.

These principles provide the basis for requirements and measures for the protection of people and the environment against radiation risks and for the safety of facilities and activities that give rise to radiation risks, including, nuclear installations, the use of radiation and radioactive sources, the transport of radioactive material and the management of radioactive waste.

## 3.2. LEGAL AND REGULATORY FRAMEWORK

The governmental and legal framework for safety establishes the roles and responsibilities of the government and the regulatory body. Paragraph 2.2 of GSR Part 1 (Rev. 1) [6] states:

> "The government establishes national policy for safety by means of different instruments, statutes and laws... For example, the government establishes laws and adopts policies pertaining to safety, whereas the regulatory body develops strategies and promulgates regulations in implementation of such laws and policies."

Paragraph 4.3 of GSR Part 1 (Rev. 1) [6] states:

> "The regulatory process shall provide a high degree of confidence, until the release of facilities and activities from regulatory control, that:
>
> (a) Safety is optimized, the balance between operational benefits and potential consequences for people and the environment being taken into account."

## 3.3. LICENSING PROCESS FOR SITING

The licensing process for a nuclear installation consists of several steps, including licensing for siting, design, construction, commissioning, operation, decommissioning and release from regulatory control. Some steps of the licensing process may be combined to facilitate the regulatory process, depending on national practice.

The regulatory body develops regulations for the licensing of nuclear installations, and accompanying guidelines to provide applicants with clarity regarding the licensing process.

The applicant develops siting procedures within its management system and the procedures are assessed, reviewed and revised periodically, taking into account operating experience and modifications, as well as lessons learned from other Member States.

Paragraph 1.15 of IAEA Safety Standards Series No. SSR-1, Site Evaluation for Nuclear Installations [5] states:

> "The siting process for a nuclear installation is divided into two stages:
>
> (a) Site survey, in which candidate sites are identified after the investigation of a large region and the rejection of unsuitable sites;
> (b) Site selection, in which the candidate sites are assessed by screening, evaluation, comparison and ranking on the basis of safety and other considerations to select one or more preferred candidate sites."

### 3.3.1. Review and assessment by the regulatory body

Requirement 25 of GSR Part 1 (Rev. 1) [6] states:

> **"The regulatory body shall review and assess relevant information — whether submitted by the authorized party or the vendor, compiled by the regulatory body, or obtained from elsewhere — to determine whether facilities and activities comply with regulatory requirements and the conditions specified in the authorization. This review and assessment of information shall be performed prior to authorization and again over the lifetime of the facility or the duration of the activity, as specified in regulations promulgated by the regulatory body or in the authorization."**

The basic objective of the review and assessment is to determine whether the licensee's application documents demonstrate compliance with all the safety requirements specified by the regulatory body throughout the lifetime of facilities and activities. The depth and scope of the review and assessment of the facility or activity is determined in accordance with a graded approach.

Paragraph 3.162 of IAEA Safety Standards Series No. GSG-13, Functions and Processes of the Regulatory Body [9] states (footnote omitted):

> "The review and assessment process should include the following steps:
>
> (a) Definition of the scope of the review and assessment process;
> (b) Specification of the purpose of and technical bases for the review and assessment process (these could be considered acceptance criteria for the review and assessment);
> (c) Identification of additional information, if necessary, for the review and assessment;
> (d) Performance of a step by step review and assessment procedure to determine whether the applicable safety objectives and regulatory requirements have been met for each aspect or topic;
> (e) Decisions on the acceptability of the authorized party's safety arguments or the need for further submissions;
> (f) Reporting and documentation."

The regulatory body provides internal guidance for its staff as to how to perform the review and assessment. The guidance provides support to the regulatory staff on how to check and confirm that the licensing documents meet safety objectives and criteria. Detailed guidance on specific topics for review and assessment is also provided, as necessary. Reference [22] provides an example guide for the review of an SAR.

### 3.3.2. Siting process

The process of siting and site evaluation for a nuclear installation comprises five distinct stages. The initial two stages — site survey and site selection — form the foundation of this comprehensive process.

Paragraph 2.3 of IAEA Safety Standards Series No. SSG-35, Site Survey and Site Selection for Nuclear Installations [29] states:

> "The siting process and the site evaluation process include five different stages. The siting process for a nuclear installation consists of the first two stages of these five, i.e. site survey and site selection. In the site survey stage, large regions are investigated to find potential sites and to identify one or more candidate sites. The second stage of the siting process is site selection, in which unsuitable sites are rejected and the remaining candidate sites are assessed by screening and comparing them on the basis of safety and other considerations to arrive at the preferred candidate sites."

During the site survey stage, extensive regions are carefully examined and analysed to identify potential sites that meet the necessary criteria for hosting a nuclear installation. This involves thorough exploration to assess various factors such as geological stability, proximity to population centres, environmental impact, and accessibility to the site at emergency. The objective is to cast a wide net, surveying large geographical areas to identify one or more promising candidate sites.

During the site selection stage, the focus shifts to a more detailed evaluation of the identified candidate sites. The goal is to systematically screen out unsuitable locations and narrow down the options to a select few that hold the most promise. This involves a rigorous screening process wherein each candidate site is scrutinized based on predefined safety criteria and other relevant considerations. Factors such as geological characteristics, seismic activity, hydrological conditions, land use patterns, and socio-economic factors are carefully assessed. Through this comprehensive evaluation process, the suitability of each site is weighed against established safety standards and regulatory requirements.

Ultimately, the outcome of the site selection stage is the identification of preferred candidate sites that exhibit the highest level of compatibility with the safety and operational needs of a nuclear installation. These sites emerge as the most viable options for further consideration and eventual development.

### 3.3.3. Site evaluation process

Site evaluation includes site characterization and consideration of factors associated with the safety features of the nuclear facilities and activities, as well as factors that might allow the release and dispersion of radioactive material in the environment.

The regulatory body reviews the site selection and characterization processes considering the potential impact of the nuclear installation and its activities on the environment and reviews the results of the analyses provided in the site evaluation report.

This site evaluation is reviewed and assessed throughout the entire lifetime of the nuclear installation, taking into account changes in the site characteristics, evaluation methodologies and relevant safety standards.

Paragraph 3.6 of IAEA Safety Standards Series No. SSG-12, Licensing Process for Nuclear Installations [11] states (reference omitted):

> "For a nuclear installation, following site selection, site evaluation typically involves the following stages:
>
> (1) Site characterization stage. This stage is further subdivided into:
>     (i) Site verification, in which the suitability of the site to host a nuclear installation is verified, mainly according to predefined site exclusion criteria;
>     (ii) Site confirmation, in which the characteristics of the site necessary for the purposes of analysis and detailed design are determined.
>
> (2) Pre-operational stage. Studies and investigations which begin in the site characterization stage are finalized before the start of construction. The site data obtained allow a final assessment of the simulation models used in the final design.
>
> (3) Operational stage. Appropriate safety related site evaluation review activities are carried out throughout the operating lifetime of the facility, mainly by means of monitoring and periodic safety review."

Paragraph 2.7 of SSG-35 [29] states:

> "There are three important steps that receive input from site survey, site selection and the site evaluation process before construction starts. These are:

(a) The decision regarding the suitability of the preferred site, i.e. confirmation that the site has no characteristics that would preclude the safe operation of a nuclear installation;
(b) The definition of the site related design basis parameters on the basis of the site evaluation report;
(c) The preparation of the preliminary safety analysis report or preliminary safety case which, among other things, demonstrates that the site related design basis parameters have been appropriately taken into account, in particular through the design features of the nuclear installation and the measures to be taken for site protection."

The site evaluation includes activities such as inspection, testing, and verification and validation to check that the results meet the specified acceptance criteria. The results of studies and investigations of the site evaluation are documented, and an independent review is performed.

During the site evaluation, an appropriate quality assurance programme in compliance with regulatory requirements is included as part of the management system. After the completion of site evaluation, the regulatory body approves the acceptability of the site before construction of the nuclear installation starts.

### 3.3.4. Environmental impact assessment

The objective of the EIA is to identify and to assess all the environmental impacts of the nuclear installation on environmental and public health.

The EIA includes the environmental conditions, natural and human induced risks and hazards for the site, and population density. It also covers environmental monitoring and plans for necessary mitigatory measures if permitted limits are exceeded.

The EIA is performed before commissioning of the nuclear installation, and the regulatory body reviews and assesses it against established regulatory criteria and in consultation with other regulatory bodies such as the environmental regulatory body.

IAEA Safety Standards Series No. GSG-10, Prospective Radiological Environmental Impact Assessment for Facilities and Activities [24] provides recommendations on the regulatory framework and assessment methodology for the radiological EIA. A non-radiological EIA is generally also conducted in accordance with the regulations of the Member State. However, as there is some overlap between the input data for both the radiological and non-radiological EIAs, good coordination between the two is advisable.

### 3.3.5. Site evaluation in the safety analysis report

The site evaluation in the SAR provides site information over the lifetime of the plant. Paragraph 3.2.4 of IAEA Safety Standards Series No. SSG-61, Format and Content of the Safety Analysis Report for Nuclear Power Plants [30] states:

"This information includes the following:

(a) The collection of site reference data for the plant design (e.g. geological, seismological, geotechnical, volcanic, hydrological and meteorological);
(b) The site-specific hazard evaluation for external events of natural origin (e.g. earthquakes, meteorological events, flooding, geological and volcanic hazards, hazards from biological organisms, surface deformation relating to tectonic (i.e. faulting) and non-tectonic causes) and of human induced origin (e.g. aircraft crashes, chemical explosions from activities performed at nearby industrial facilities and other facilities);
(c) The design targets in terms of the probability of recurrence of external events, with account taken of their severity and associated uncertainties;

(d) An evaluation of the impact of the site related issues to be considered in the parts of the safety analysis report on emergency preparedness and response and accident management;
(e) The arrangements for the monitoring of site related parameters throughout the lifetime of the plant;
(f) The potential for specific hazards to give rise to impacts simultaneously on several units in the case of a multiple unit site."

Further details on specific areas of site evaluation such as external human induced events and geotechnical aspects are described in Refs [23, 29].

## 3.4. INSPECTION BY THE REGULATORY BODY

Requirement 27 of GSR Part 1 (Rev. 1) [6] states that "**The regulatory body shall carry out inspections of facilities and activities to verify that the authorized party is in compliance with the regulatory requirements and with the conditions specified in the authorization.**" The regulatory body verifies the contents of the documents submitted by the applicant during the inspection phase and confirms the information and data necessary for review and assessment.

Paragraph 4.52 of GSR Part 1 (Rev. 1) [6] states that "Regulatory inspections shall cover all areas of responsibility of the regulatory body, and the regulatory body shall have the authority to carry out independent inspections."

Paragraph 3.225 of GSG-13 [9] states:

> "The regulatory inspection programme should be comprehensive and consistent with the overall regulatory strategy. The inspection programme should be thorough enough to ensure that the regulatory objectives and requirements are being met, thereby providing the regulatory body with a confidence that the authorized party is effectively maintaining the safety of the facility or activity. The inspection programme should also be developed so that the regulatory body can determine whether the authorized party conducts activities in accordance with previously established procedures, and has an effective self-assessment process capable of prompt identification and correction of actual and potential problems."

The priority and frequency of inspections are based on the risk associated with the radiation source and the complexity of the facility or activity, in accordance with the graded approach.

Recommendations on the aspects to be confirmed in the regulatory inspection are provided in SSG-12 [11]; for example, para. 3.9 of SSG-12 [11] provides recommendations on site acceptability and para. 3.39 provides recommendations on the conditions to be fulfilled for a construction licence.

### 3.4.1. Internal guidance on inspection

Paragraph 3.262 of GSG-13 [9] states:

> "The regulatory body should issue internal guidance for its inspectors on performing regulatory inspections in order to ensure a consistent approach to inspection while allowing sufficient flexibility for inspectors to take the initiative in dealing with new concerns that arise. Each inspector should be given adequate training in following this guidance."

Paragraph 3.263 of GSG-13 [9] states:

> "The guidance for inspectors should include the following:
>
> (a) Policies of the regulatory body regarding inspections.
> (b) The legal basis for regulatory inspection and the scope of the inspector's authority.
> (c) The use of regulatory requirements, regulations, guides and standards.

(d) The development of an inspection programme.
(e) The implementation of the inspection programme, including:
   (i) Facilities (or areas of the facility) or activities to be subject to inspection;
   (ii) Method of inspection to be used;
   (iii) Methods for selection of inspection samples;
   (iv) Use of relevant technical information;
   (v) Use of inspection questionnaires;
   (vi) Follow-up on inspection findings.
(f) Reporting requirements and practices for inspectors.
(g) Standards of conduct of inspectors.
(h) The enforcement policy, procedures and practices."

#### 3.4.2. Inspection procedures

Paragraph 3.268 of GSG-13 [9] states:

"The inspection procedures of the regulatory body should incorporate and use a variety of methods, as follows:

(a) Monitoring and direct observation (such as of working practices and equipment);
(b) Discussions and interviews with personnel of the authorized party and of the contractor, if necessary;
(c) Examination of procedures, records and documentation;
(d) Confirmatory tests and measurements."

Examples of an inspection procedure and guidelines for the preparation and performance of an inspection are given in Annex III.

#### 3.4.3. Site inspection

Before the construction of a nuclear installation begins, the regulatory body reviews, inspects and monitors the site preparation activities and verification results of site characterization by the operating organization.

The regulatory inspection verifies that the operating organization's siting activities are in compliance with regulatory requirements and acceptance criteria. During site preparation, the regulatory body also confirms that the site characteristics are consistent with the description in the documents submitted for the siting licence.

### 3.5. ENFORCEMENT BY THE REGULATORY BODY

Requirement 30 of GSR Part 1 (Rev. 1) [6] states that "**The regulatory body shall establish and implement an enforcement policy within the legal framework for responding to non-compliance by authorized parties with regulatory requirements or with any conditions specified in the authorization.**"

Regulatory enforcement actions by the regulatory body address non-compliance by the licensee with specified conditions and requirements. These actions require the licensee to correct relevant procedures and practices or, when necessary, to change or improve the design of structures, systems and components important to safety.

# 4. IAEA CAPACITY BUILDING PROGRAMME FOR THE REVIEW OF NUCLEAR INSTALLATION SITE SAFETY

The IAEA's capacity building programme for the review of nuclear installation site safety is designed to cater to the specific educational and practical needs of regulatory bodies and affiliated entities within Member States, such as utility organizations, technical support organizations (TSOs) and nuclear energy programme implementing organizations (NEPIOs). This programme is strategically designed to strengthen the competency and capability of managers and technical personnel engaged in site and design safety for nuclear installations, particularly in relation to the challenges posed by external hazards. The training consists of both in-person sessions and e-learning.

The implementation of the IAEA capacity building programme in a Member State is activated through an official request by a governmental body, referred to hereafter as the host organization. The programme begins with the host organization's self-assessment to identify specific areas for training. The IAEA then formulates a bespoke capacity building training programme on the basis of the self-assessment results.

During the preparatory phase, trainees from the host organization undertake preliminary self-study through the IAEA's e-learning hub. This phase equips participants with essential knowledge pertinent to the upcoming training course. In collaboration with a designated contact person from the host organization, the IAEA assigns a technical officer to this training mission, referred to as a 'mission team leader'. This team leader then designs the main training curriculum and oversees the implementation of the training mission.

The central facet of the capacity building programme is the execution of training course, which can encompass a combination of traditional in-person lectures and immersive online e-learning that amalgamate theoretical knowledge and practical applications. At the end of the training course, the IAEA produces a comprehensive mission report, detailing the materials used and the course achievements.

Central to the efficacy of the capacity building programme is the incorporation of a strong feedback system. Following the conclusion of the training, the host organization is strongly urged to contribute insights, reflections and suggestions, thereby cultivating an environment of ongoing enhancement. This feedback helps to improve the training in the future, ensuring a forward-moving and enduring method for fostering capacity building.

The capacity building programme for the review of nuclear installation site safety reflects the commitment of the IAEA and its Member States to increase expertise and preparedness in the realm of nuclear installation site safety. This synergistic collaboration among regulatory bodies, governmental organizations and the IAEA embodies a complete and adaptive approach towards ensuring the safety of nuclear installations against external hazards. A schematic flowchart showing the typical process followed to implement the IAEA capacity building programme is presented in Table 1. Further information about each step in the process is provided in the following subsections.

## 4.1. SELF-ASSESSMENT AND ESTABLISHMENT OF CAPACITY BUILDING ACTION PLAN

Self-assessment is a valuable tool where there is a lack of data regarding the competency of the host organization and its education and training beneficiaries. The self-assessment can be conducted internally or by an external organization such as the IAEA. It offers insights into the technical capabilities of managers and technical staff, enabling the organization to systematically pinpoint and address areas for enhancement.

TABLE 1. A TYPICAL STEPWISE PROCESS FOR IMPLEMENTATION OF THE IAEA CAPACITY BUILDING PROGRAMME

| IAEA | HOST ORGANIZATION |
|---|---|
| **Initiation** | |
| | 1. Request for a training mission |
| 2. Review and acceptance of the request | |
| 3. Designation of mission team leader | 4. Designation of contact person |
| 5. **Virtual Meeting** for instructions on the mission process | |
| **Self-assessment and establishment of capacity building action plan (CBAP)** | |
| 6. Preparation and provision of self-assessment package | 7. Performance of self-assessment |
| 8. Evaluation of self-assessment results | |
| 9. Preparation of draft CBAP | |
| 10. **Virtual Meeting** for discussion and establishment of CBAP | |
| **Design of training course** | |
| 11. Establishment of terms of reference (ToR) | 12. Review and agreement of ToR |
| 13. Composition of mission team | 14. Submission of a preliminary list of trainees |
| 15. Development of training programme agenda | |
| 16.**Virtual Meeting** for discussion and agreement of draft agenda and necessary arrangements | |
| **Self-study through e-learning platform** | |
| 17. Provision of access to IAEA e-learning platform | 18. Self-study using IAEA e-learning platform |
| **Preparation for the training mission** | |
| 19. Preparation of training materials | 20. Preparation of national presentation |
| 21. Sharing training materials | 22. Preview of training materials |
| 23. **Briefing meeting** of IAEA mission team | |
| **Implementation of training course** | |
| 24. Opening and national presentation (Module 1) | |
| 25. Lectures on general aspects (Module 2) | |
| 26. Lectures on country-specific technical aspects (Module 3) | |
| 27. Hands-on practice (Module 4) | |
| 28. Feedback to participants and conclusion (Module 5) | |
| **Evaluation and feedback** | |
| 29. Evaluation of the mission and feedback | |
| **Post-training activities** | |
| 30. Preparation of the mission report | 31. Review of the mission report |
| 31. Confirmation of the mission report | |
| 32. Delivery of final mission report to the participants | |

Analysing the self-assessment results allows the strengths and weaknesses in the host organization's capacities to be identified. This provides a baseline for IAEA Site and External Events Design (SEED) mission teams to establish a proper capacity building action plan (CBAP) for the host organization and/or the Member State as a whole. Accurately recorded self-assessment results and a well-designed CBAP lead to a successful capacity building programme and enhanced capabilities of the managers and technical staff working in the areas of the review of nuclear installation site and design safety against external events in Member States.

### 4.1.1. Self-assessment

Self-assessment allows the host organization to proactively engage, and subsequently submit the outcomes to the IAEA SEED mission team. If a self-assessment method has not been developed internally or is deemed insufficient, the host organization may also choose to utilize self-assessment tools provided by the IAEA. The choice of approach will be determined through discussions with the IAEA mission team to establish the optimal assessment methodology.

Self-assessment involves a carefully constructed set of question that, encourage interaction. Following a decision to perform the self-assessment, the host organization may find it useful to establish two distinct working teams. One team, consisting of technical staff, prepares a draft response to the questionnaire, while another team, made up of senior staff and managers, reviews the draft. These teams collaborate harmoniously by discussing and refining these responses to precisely represent the current status of the organization. To ensure a thorough self-assessment, a time frame of approximately one month is needed.

In the case of the self-assessment tool provided by the IAEA, the host organization answers approximately 500 questions across 53 topics under 5 major categories, developed based on the IAEA safety standards relating to nuclear installation site safety. The responses reflect the competence and confidence levels of staff in the host organization.

The contact person of the host organization submits the self-assessment results to the IAEA SEED mission team leader who, along with the mission team, evaluates the results to derive the capacity and competency levels of the host organization or of the Member State as a whole.

The five areas of the IAEA self-assessment tool are as follows:

(a) Management system;
(b) Site selection;
(c) Site evaluation;
(d) Regulatory documents and framework;
(e) Monitoring and periodic review.

The 53 topics covered within these five areas are fundamentally based on the requirements and recommendations outlined in the IAEA safety standards relating to nuclear installation site safety. A technical analysis of each of the 53 topics yields insights into the strengths and vulnerabilities of particular technical domains, including seismic and flooding hazards, radiological EIA and human-induced events.

The evaluation of the host organization's responses defines areas for improvement in relation to the IAEA safety standards, ultimately providing the fundamental information necessary for developing tailored and targeted capacity building training plans for the organization to address these gaps.

### 4.1.2. Establishment of capacity building action plan

The IAEA SEED mission team leader, in collaboration with the team members, establishes a tailored CBAP for areas of vulnerability within the host organization, based on the strengths and weaknesses identified through self-assessment across different fields. The CBAP encompasses concise education and training programmes for specific vulnerabilities and topics. Recommendations or suggestions identified through previous IAEA missions also need to be considered in this process.

There are two main training methods available: online learning via the e-learning platform for self-directed study, and offline approaches like classroom training and workshops. Online learning encompasses essential content related to the safety of nuclear installations, incorporating IAEA safety standards requirements and recommendations. Conversely, offline education delivers tailor-made training that aligns with the host organization's needs, conducted within the respective country.

In general, each topical area such as seismic hazard assessment, radiological dispersion, human-induced hazards assessment (often covered by a single IAEA safety guide), is covered by one training course. However, it would also be possible for one course to cover two topical areas that are related to each other, for example, flooding hazard assessment caused by meteorological, hydrological and human-induced events, or site characteristics important to radiological EIA and emergency planning.

In cases where shared vulnerabilities are identified through self-assessment across multiple Member States, however, opting for a regional or interregional training course might prove a more efficient and effective approach. Such training could be hosted in one of the affected Member States or at the IAEA headquarters.

The CBAP, drafted by the IAEA, is optimized and confirmed through review and discussion with the host organization. This process may include a virtual meeting between representatives from the host organization and the IAEA.

## 4.2. DESIGN OF TRAINING COURSE

### 4.2.1. Establishment of terms of reference

In tandem with the contact person from the host organization, the IAEA mission team leader formulates the terms of reference for the training course. These terms of reference commonly encompass background, goals, scope, time frame, schedule, hosting institution, location, target trainees, the mission team leader, technical expertise expected from the mission team members, preparatory arrangements, reference materials, guidelines for conduct, and envisioned outcomes of the training course. The applicable IAEA safety standards and other resources that pertain to the training course are included as supplementary reference materials.

The terms of reference are periodically adjusted as necessary in the run up to the training course, with the agreement of the IAEA and the host organization.

### 4.2.2. Composition of mission team and trainees

Once the terms of reference have been endorsed by both the IAEA and the host organization, the mission team leader recruits the mission team members who possess the expertise necessary to comprehensively address the specified technical areas.

The recruitment procedure involves the selection of professionals with the requisite subject knowledge and a demonstrated track record in their respective fields, which ensures the goals and objectives of the training course are attained. The mission team leader aims to select team members who have experience in various nuclear siting projects around the world, from which they can bring practical insights, thus enhancing the quality of the course. The mission team leader provides the team members with the terms of reference and the host organization's self-assessment so that they might understand the needs of the future trainees and the depth of expertise needed in the specific technical domain.

At the same time, the contact person from the host organization is responsible for selecting trainees for the training course. Given the close collaboration among several organizations involved in a nuclear programme, such as NEPIOs, regulatory bodies, utilities and TSOs, especially as regards safety reviews and licensing processes, it is advantageous, efficient and synergistic to include participants in the training course from a range of roles within these organizations. This inclusive approach encompasses both managerial and technical levels and extends across various functions within the Member State.

Once the mission team members and trainees have been established, the mission team leader acquaints them with the e-learning platform. This platform will facilitate their preliminary study of the fundamental knowledge needed for training on the safety review of nuclear installation sites.

### 4.2.3. Development of the training course agenda

The mission team leader develops a draft training course agenda together with the team members, according to the terms of reference. The mission team leader finalizes the draft agenda through consultation with the host organization contact person.

The mission team leader designs the overall structure of the training course and presents it in a simplified table. After agreement on the overall structure with the host organization, the mission team leader develops a detailed draft agenda, consulting with the presenters (i.e. the mission team members and the national presenters).

Optimal results are achieved by structuring the programme in several sessions a day, each one lasting 1.5–2 hours and encompassing approximately two activities. The provision of breaks for coffee and lunch is also beneficial. This structure ensures that participants remain fully engaged and attentive, maximizing their absorption of crucial concepts and skills.

## 4.3. SELF-STUDY THROUGH THE E-LEARNING PLATFORM

The NUCLEUS information repository, overseen by the IAEA, provides a diverse array of scientific, technical, and regulatory assets that include databases, applications, publications and training materials. Within the framework of NUCLEUS, the EESS has established a knowledge and resource platform referred to as the EESS Hub[1]. This hub is available to registered members from various Member States, offering an overview of the EESS and providing news and updates on ongoing projects, relevant resources, publications, webinars, technical events, and the e-learning platform.

The e-learning platform delivers education and training services independent of other IAEA services. This platform is designed to assist individuals from Member States in understanding the structure and content of IAEA safety standards and provide general information concerning nuclear installation site and design safety against external events. The training materials are user-friendly and enable participants to engage in self-guided study at their preferred pace.

The e-learning platform can be used by host organizations prior to planned training courses or workshops, as preparatory training. Furthermore, it can be used by members of IAEA mission teams, especially those participating for the first time, to familiarize themselves with the relevant IAEA safety standards and general disciplinary knowledge within the training course's scope.

### 4.3.1. Self-training implementation

The e-learning platform provides comprehensive training material comprising sequential steps, encompassing self-assessment and self-study employing standardized lecture presentations and e-learning modules, hands-on practice in SAR reviews, and a feedback questionnaire for the service. Ideally, all the training modules will be completed in sequence to optimize subsequent training activities, but the option to undertake individual modules is also available. The following sections elaborate on the details of these steps.

#### *4.3.1.1. Self-assessment*

The self-assessment feature on the e-learning platform is designed to assess the trainee's knowledge of, and confidence in implementing, IAEA safety standards relating to the review of nuclear installation site and design safety. Two types of self-assessment questionnaire exist: one for organizations and another for individuals. The questionnaire for organizations evaluates regulatory bodies, utilities, NEPIOs, and TSOs to assess their level of confidence in adhering to IAEA safety standards. The second questionnaire is directed at managers and personnel working within organizations, as well as individuals within the

[1] https://gnssn.iaea.org/main/EESS%20Hub/Pages/default.aspx

field of nuclear safety, aimed at assessing their grasp of IAEA safety standards on nuclear installation site and design safety against external events. Upon submission of the online self-assessment form, the responses are automatically evaluated.

If an organization wishes to undertake an official self-assessment and receive feedback from the IAEA, the designated contact person from the organization initiates a request with the EESS Section Head through official IAEA channels, as outlined in Section 4.2.

#### *4.3.1.2. Self-study on the IAEA safety standards*

On the basis of the results of the self-assessment, trainees select pertinent self-study modules. If a training course is on the horizon, the EESS may assign specific modules for completion. The self-study materials accessible through the e-learning platform are rooted in IAEA safety standards linked to site and design safety against external hazards. The training materials encompass a blend of course overviews, e-learning modules, standard lecture materials, e-books, and links to pertinent information sources. Presently, lecture materials are offered for three domains as listed below:

— Site Survey and Site Selection for Nuclear Installations (SSG-35) [29];
— Potential Effects of Nuclear Installations on the Region (SSR-1) [5] and (NS-G-3.2) [31];
— Site Evaluation for Nuclear Installations (SSR-1) [5].

Upon completing each self-study module, trainees are encouraged to partake in a post-training quiz to self-assess their comprehension and progress. The quiz content aligns with the training content, and trainees' responses are scored. The quiz outcomes might be used internally to enhance the quality of e-learning materials. The content may undergo updates based on feedback from Member States and trainees. Further specifics and updates on training courses and materials are accessible via the e-learning platform.

#### *4.3.1.3. Practice review of safety analysis reports*

The SAR review practice module serves as a self-training resource, designed for a group of individuals to comprehend the SAR review process, and engage in a simulation of the review using a comprehensive set of sample documents. This course aims to provide future reviewers with experience of safety review activities within a team-based practice setting.

Participants gain the opportunity to appreciate diverse perspectives, including those of regulatory bodies, utilities, and local residents, via role-playing activities within the context of review of site safety for nuclear installations. Each team acts as a regulatory body when it reviews the SAR, raises RAIs, and prepares the SER, while each team acts as a utility when it develops answers to the RAIs from other team(s).

The SAR review process is described, detailing its various stages. The process begins with the formation of teams: normally one moderator/operator and a minimum of two teams, each consisting of a leader and two or more staff members. The teams identify regulatory safety aspects and raise RAIs, which are then presented along with a Q&A session for clarification. Comprehensive responses to the RAIs are developed, presented, and further clarified through a Q&A session with the team that issued the RAIs. Finally, a SER is prepared and presented, accompanied by a dedicated Q&A session to answer questions from the public.

This course accommodates individuals aiming to gain an understanding of the SAR review process. Additionally, it serves as a resource for trainees partaking in capacity building training courses and workshops, providing them with supplementary SAR review practice. All the necessary materials for the practice are available in the e-learning platform, such as instructions, SAR review guidelines, sample SARs and SERs, and templates for RAIs.

#### 4.3.2. Training evaluation questionnaire

Upon successfully completing all assigned self-study modules, trainees are invited to provide feedback on their content and structure through a questionnaire. The questionnaire results are compiled and utilized internally to continually enhance the calibre of the training programme. Those who voluntarily engage in the modules are also encouraged to participate in this questionnaire.

### 4.4. PREPARATION FOR THE TRAINING MISSION

#### 4.4.1. Preparation of training materials

The SEED mission team leader needs to engage with all members of the mission team as well as other presenters to ensure the consistent and systematic development of lecture materials and hands-on practice materials for the capacity building training course.

The mission team leader provides presenters in advance with the latest IAEA presentation template, and informs them of the importance of carefully reviewing the content and desired outcomes. The presentation template is structured to include essential components such as an overview, up-to-date IAEA safety standards, key concepts, assessment methodologies, examples, case studies, and references. Clear instructions are given to presenters where necessary. Furthermore, a critical role of the IAEA mission team leader is to verify whether the lecture materials include relevant real-life cases and case studies that align with the respective criteria of IAEA safety standards. This systematic and content-rich approach enhances the training experience, fostering greater engagement and improving the overall effectiveness of the education of the trainees.

The mission team leader asks presenters to ensure that key messages are highlighted in their lecture materials, to enable trainees to concentrate on these essential points. High quality lecture materials present key information and keywords at a glance, even if they include numerous images, charts, and text that need to be revisited for further learning after the lecture. Skilled educators consistently emphasize crucial messages and keywords, providing context throughout the lecture. This is because, in reality, trainees only retain a portion of the information provided in the lecture, and are likely to remember even less when undertaking tasks outside the classroom.

Ideally, the lecture lasts around 70–80% of the allotted presentation time, with the remaining time allocated to Q&A. Providing sufficient time for Q&A helps both the instructor and the trainees to focus more on the essential points, enables a focus on topics that trainees are genuinely curious about, and encourages active participation from both the instructor and the trainees.

The mission team leader also requests team members to create quizzes based on the lecture content. The purpose of these quizzes is to assess the trainees' level of understanding of the lecture. Around five questions per lecture are created, in general, ranging from simple ones that can be easily answered based on facts, to more challenging ones that involve comprehensive application of the learning material. Both pre-training and post-training quizzes can be assigned to the trainees; by comparing the results of both, the extent of improvement in the trainees' understanding and knowledge can be ascertained.

#### 4.4.2. Sharing and preview of training materials

The mission team leader creates a web page for the training course under the EESS Hub and uploads training materials there as they become available. The training course participants are expected to preview the training materials before the training course and bring to the course questions and/or topics for further discussion. The preview of training materials enhances the outcomes of training and leads to more active engagement in the training activities. Therefore, the mission team leader ensures all training materials are shared on the web page well before the training course to allow participants sufficient time to familiarize themselves with the content.

For hands-on exercises like practice reviews (see Section 4.3.1.3), the sharing of sample documents such as the SAR and SER is essential, as these documents carry crucial insights into safety assessment and evaluation procedures. Given the complexity of these documents and the time needed to comprehend them, it is advisable to share them even earlier than the regular preparatory period. This gives participants sufficient time to explore all the sections, investigate specific topics, and seek additional information as necessary. This early engagement enhances their preparedness and active participation during the training itself.

#### 4.4.3. Other arrangements for the training course

Working closely with the contact person from the host organization and IAEA staff, the mission team leader identifies crucial elements and plans needed for the successful execution of the training course. These arrangements go beyond what is outlined in the terms of reference. They encompass various aspects such as making financial preparations, compiling the documentation for visa applications, coordinating mission team members' travel, securing suitable workshop space (including appropriate rooms with presentation equipment), sending invitations to participants, managing the participants' list, and managing general logistical aspects.

A preparatory virtual meeting between the mission team and the representatives of the host organization(s) may be useful. The agenda of the meeting may include confirmation of the draft training course agenda, participants' list, workshop materials and the training course web page, venue, transportation arrangements for the mission team, and other issues as necessary.

#### 4.4.4. Mission team briefing meeting

The mission team leader and members have a team briefing meeting just before the training course to confirm that all the necessary arrangements are in place.

The mission team leader clarifies roles and responsibilities, considering the distinct technical domains of each member, and runs through the schedule of presentations in the training course. As a result, the team briefing meeting establishes a cohesive foundation for the subsequent stages of the training, promoting a unified understanding and improving execution of the training course's mission.

The team briefing meeting is generally held either in person on the arrival day at the designated duty station, or virtually the week before the course, depending on the circumstances.

## 4.5. IMPLEMENTATION OF THE TRAINING COURSE

The training course has a standard duration of five days and comprises five distinct modules, as detailed in subsections 4.5.1–4.5.5. However, the mission team leader and host organization have the flexibility to tailor the modules according to specific needs. Examples of the overall structure and agenda of the training course are provided in Section I–4, Annex I.

The training course agenda designates time for discussions, fostering engagement and active participation among participants. To gauge and enhance trainees' comprehension, quizzes are administered both before and after lectures, offering valuable insights into their evolving understanding of the subject matter.

#### 4.5.1. Opening and national presentation (Module 1)

The initial module of the training commences with the opening session, at which attendees are greeted with addresses from a high-level manager representing the host organization. This session involves further introductions, allowing participants to get to know each other. Moreover, a comprehensive overview of the training's objectives and the ensuing agenda is outlined by the mission team leader.

Subsequently, the national presentation sheds light on pivotal aspects of the host State's nuclear power programme. This encompasses the regulatory framework, procedural underpinnings, and nuclear installation site and design safety in the face of external events.

The national presentation provides an opportunity to share critical knowledge and insights, establishing a foundation for mutual comprehension and collective discourse among attendees. This module thus sets the stage for an enriched and collaborative learning experience throughout the training.

#### 4.5.2. Lectures on general aspects (Module 2)

Prior to commencing the lectures, a preliminary set of quizzes assesses participants' baseline understanding of general and technical topics.

The sequence of lectures, presented by the mission team, encompasses comprehensive aspects related to site and design safety against external events at nuclear installations. Topics covered may include:

- Recent IAEA activities in site and design safety against external events;
- IAEA safety standards and supporting documents overview;
- Safety considerations for nuclear installations against external events;
- Licensing processes and safety reviews;
- Regulatory capabilities in site and design safety against external events.

The intensity of Module 2 may be modulated based on the participants' overall knowledge, or supplemented with online self-study or e-learning modules accessible through the EESS Hub.

#### 4.5.3. Lectures on country-specific technical aspects (Module 3)

The mission team gives a series of lectures focused on country-specific technical subjects, highlighting areas where the enhancement of capacity is particularly warranted for participating managers and technical staff.

Each lecture focuses on a distinct area of expertise, encompassing a range of important topics, including a thorough exploration of relevant IAEA safety standards, essential principles that govern the subject matter, methodologies, and overarching concepts, as well as case studies that bring the theoretical framework to life. This multifaceted approach ensures that participants gain a deep and nuanced understanding of the topic at hand.

The incorporation of sample documents, such as the sample SAR for hands-on practice, for the lectures on case studies enriches the learning experience. By utilizing these practical resources during case study discussions, participants can witness how theoretical knowledge translates into real-world applications.

Upon the culmination of this series of comprehensive technical lectures, the same set of quizzes given at the start of the lectures is once again presented to measure the extent of improvement in the participants' comprehension of the subject matter.

#### 4.5.4. Hands-on practice (Module 4)

An entire module of the training is dedicated to hands-on practice, offering participants the opportunity to gain practical insights in a specific area. This segment bridges the gap between theory and application, allowing participants to engage in experiential learning. The chosen practice topics align with the technical subjects covered in the preceding lectures.

The mission team leader ensures that participants are well prepared for the practical exercises, providing them with essential materials, practical guidelines, sample documents and computer codes in advance.

Participants are also encouraged to bring their own equipment or tools for this module, such as laptops with pre-installed computer codes and datasets. For practical exercises involving team-based activities, the mission team leader collaborates with the host organization's contact person to develop a preliminary plan for team composition outlining individual roles within each team. The communication of these roles and tasks is facilitated by team leaders, ensuring a coordinated and structured practice session, as per the guidelines provided by the mission team leader.

Where applicable and feasible, this module may encompass a visit to a potential national site for a nuclear installation. This adds a practical dimension to the learning experience, allowing participants to interact with real-world scenarios. However, the feasibility of such a visit is contingent on the availability of a site and agreement by the host organization.

### 4.5.5. Feedback to participants and conclusion (Module 5)

The training course ends with a dedicated feedback and conclusion module. The mission team leader, with support from other team members, presents a comprehensive summary of the training course, including the activities undertaken, an evaluation of the modules, and an assessment of the pre- and post-training quiz outcomes.

Finally, the closing session involves discussions about upcoming activities, a potential certificate presentation ceremony (as per the host organization's request), and concluding remarks from both the mission team leader and the host organization's management. The involvement of high-level and senior management from the host organization allows them to hear the training course's summary, grasp critical national concerns, and discern the future steps to enhance site and design safety for nuclear installations in the face of external events.

## 4.6. EVALUATION AND FEEDBACK

During or after the final module, the mission team leader invites training course participants to contribute their insights through a training course evaluation questionnaire . This questionnaire serves as a valuable tool for gathering feedback on the training programme's effectiveness and participant satisfaction. The collected results play a pivotal role in refining and optimizing future iterations of the programme's content and structure.

Templates for the training course evaluation questionnaire can be found in Annex I.

## 4.7. POST-TRAINING ACTIVITIES

Upon completion of the capacity building training course, the mission team leader compiles the outcomes of the event into an IAEA mission report. The mission report serves as a source of information for technical officers who undertake future missions or similar IAEA programmes in that country. The mission team leader classifies this information according to its intended audience and shares or stores it through suitable channels.

The mission report is based on facts, outlining the main activities and conclusions of the training course. After the report has been reviewed by EESS and the host organization, the mission team leader finalizes it and shares it on the training course web page within two months, ensuring participants have access to resources for further study. The mission team leader also uploads any updated training materials on the web page. A sample outline of the elements typically presented in a training course summary report, intended for distribution among participants, is given in Annex I.

This mission report and subsidiary materials are stored in EESS's internal repository, too, for the staff of the EESS to use them as reference for future training activities.

# 5. CAPACITY BUILDING PROGRAMMES FOR THE REVIEW OF NUCLEAR INSTALLATION SITE SAFETY WITHIN THE REGULATORY BODIES OF MEMBER STATES

This section provides several examples of capacity building programmes that have been successfully implemented in several Member States. There are many commonalities among the programmes, including their objectives and the general subject matter. However, each Member State tailors its own capacity building programme to its specific circumstances, particularly following its self-assessment and gap analysis, as discussed in Section 2.2.4.

## 5.1. REPUBLIC OF KOREA – KOREA INSTITUTE OF NUCLEAR SAFETY

The Korea Institute of Nuclear Safety (KINS) established a career development programme as a tool for determining career options, education and experience opportunities, and competency requirements for employees. A competency-based human resource management programme supports the career development programme and is concentrated on developing and implementing the competencies of individuals, teams (or groups) and the organization.

The human resource management programme assesses and identifies the current capabilities of KINS staff, and focuses on developing the competencies where they are lacking. The programme strengthens the regulatory expertise of KINS in a comprehensive and systematic way, assisting with knowledge acquisition, transfer, sharing and retention.

The competency based human resource management programme focuses on having the right people with the right knowledge, skills and attitudes in every specific professional area, and on transferring the experience and knowledge of the retiring generation to the next generation.

### 5.1.1. Organization and human resources

KINS established its International Nuclear Safety School (INSS) as a national and regional capacity building centre in 2008. The objective and role of the INSS is to develop and run training courses for regulatory staff of the Republic of Korea and other countries, and to share expertise with embarking countries.

KINS signed a memorandum of understanding with the IAEA in January 2008, to provide IAEA training courses to third parties and to further facilitate cooperation in education and training programmes with the IAEA.

### 5.1.2. Capacity building training programme

#### *5.1.2.1. General training programme of the International Nuclear Safety School*

The capacity building programme at INSS consists of four areas as follows:

(a) Leadership development programme

To enhance the organizational culture in performing the mission goals and strategies. Courses are provided for executives, managers (or candidates) and other employees.

(b) Management issues sharing programme

To share strategic management issues with all the employees to cope with the future challenges from social to industrial changes.

(c) Common competencies

To improve job efficiency by means of enhancing basic working capacity. This includes courses such as planning, creative thinking, interpersonal relationships, communication, problem solving and project management.

(d) Functional competencies programme

Includes courses on budget, finance, administration, IT skills, language, and so on.

In particular, a regulatory competence programme has been developed to ensure that legally qualified personnel perform nuclear safety regulatory activities and to ensure that they secure and maintain the expertise required by the relevant laws. There are three mandatory courses as follows:

— Radiation protection training course;
— Emergency preparedness training course;
— Nuclear regulatory inspectors training course.

For the nuclear regulatory inspectors training course, there are four areas of inspector certification available: (a) Facility management, (b) Quality assurance, (c) Radiation management, and (d) Radiological emergency preparedness. These are available both for new inspectors and as refresher courses for existing inspectors. After completion of the new inspector training, trainees undertake two years of apprenticeship in regulatory inspection activities to learn from their seniors.

The education and training programmes by competency system is summarized in Table 2.

TABLE 2. EDUCATION AND TRAINING PROGRAMMES BY COMPETENCY AT KINS

| COMPETENCY | | EDUCATION AND TRAINING PROGRAMMES |
|---|---|---|
| Job competency | Behavioural skill | New employee training, regulatory system, management system, safety culture |
| | | Management assessment, stakeholder relationship, communication, media training |
| | Regulatory professional | Inspector qualification (basic / certification / refresher) |
| | | Simulator training for regulators (basic / professional) |
| | | Regulatory professional (technical areas) |
| | Managerial professional | Regulatory introduction, regulatory policy, international cooperation |
| | | Strategic and planning, human resource management/development, budget and accounting, performance management, auditing |
| Leadership competency | New appointee | New senior, new principal |
| | By position | Senior |
| | Directors | Directors |
| Core competency | Basic | Vision and core values, IT, language, statistics |
| | Compulsory | Radiological protection, radioactivity prevention, ethics. |

There are two types of training programmes for staff in KINS: one for newly recruited staff and the other for established junior and senior staff. For the new staff, a training programme of one to three months is provided and the training areas are as follows:

- Nuclear safety fundamentals;
- NPP design;
- Laws and regulations;
- Licensing process and related documents;
- Licensing review and assessment;
- Regulatory inspection;
- Quality assurance inspection;
- Radiation protection;
- Emergency Preparedness & Planning;
- Radioactive waste management;
- APR-1400 simulator training.

Lectures to be provided to new staff include but are not limited to the following:

- Introduction of KINS;
- Nuclear safety fundamentals;
- Concept of radiation protection;
- Radiation protection;
- Safety concept and design of NPPs;
- Overview of NPP systems;
- Research reactors and nuclear fuel cycle facilities;
- Operational safety and operational guidelines of NPPs;
- Investigation of events and accidents at NPPs and root cause analysis;
- Quality assurance;
- Nuclear regulatory system;
- Laws and regulations;
- Probabilistic safety assessment and its application;
- Emergency preparedness and planning;
- Regulation of radiation sources;
- Radioactive waste management;
- Licensing documents;
- Radiation exposure management;
- Safety review and assessment;
- Regulatory inspection;
- Nuclear accidents and events;
- Attitude of regulatory bodies;
- Integrated management system;
- Nuclear safety culture;
- Analysis of radioactive material;
- Industrial and medical radiation safety;
- Environmental radiation safety.

For established junior and senior staff, it is mandatory to take intermediate and advanced courses for at least 40 hours per year. Each technical department also provides in-house training programmes annually and this training is evaluated as a departmental activity.

#### *5.1.2.2. Training programme for site evaluation*

Specialized training programmes for the staff of each technical department at KINS are generally structured in four parts. The training programme for site evaluation is as follows:

— Fundamental course:
  - Introduction;
  - Geography and demography;
  - Human-induced hazards;
  - Meteorology;
  - Hydrology;
  - Geology, Seismology and Geotechnical engineering.

— Intermediate course:
  - Introduction;
  - Geography and demography;
  - Human-induced hazards;
  - Meteorology;
  - Hydrology;
  - Geology;
  - Seismology;
  - Geotechnical engineering.

— Advanced course:
  - On-the-Job Training (A structured, classroom-style training program conducted in a one-on-one mentoring format, where senior staff guide and support new employees);
  - On-the-Job Participation (A hands-on, on-site training conducted during regulatory inspections at nuclear power plants (NPPs), utilizing the same one-on-one mentoring approach between senior and new staff).

— Special lecture course:
  - Site-specific earthquake ground motions for NPPs;
  - Recent slope stability analysis and design criteria;
  - Design criteria improvement with advance estimation method of design wave in the deep sea;
  - State-of-the-art technology of micro earthquake analysis;
  - Characteristics and application of software for general earthquake analysis;
  - Investigation of underground water at the second surface disposal facility and its modelling evaluation.

### 5.1.3. Knowledge management

Knowledge management in the area of nuclear safety can be achieved with various approaches, tools and techniques. It is important to sustain and improve the competence of both individuals and the regulatory body to use knowledge effectively and responsibly for nuclear safety regulation. Elements such as information generation, information sharing, databases, and transfer of knowledge are considered within knowledge management.

Recently, the need for replacing retiring staff has become a key challenge; therefore, nuclear knowledge management has become an increasingly important focus of management. To meet this challenge, KINS developed appropriate knowledge management methods including information and records, work processes, data interpretation, and analysis and verification techniques.

For example, when KINS staff reach a senior level (in general, after 25 to 30 years of experience in regulatory activities) they may begin to review their prior work including reports, memos, and working experiences and/or develop their own records and transfer them to their technical team or other relevant teams. They may also participate in the safety review and assessment and regulatory inspection in the licensing of NPPs as a mentor for the 1:1 mentoring program.

## 5.2. TÜRKIYE – NUCLEAR REGULATORY AUTHORITY

Rapid developments and advances in science and technology necessitate learning new information and technologies in every profession, whether in the private sector or in the public sector, and training employees in these subjects. Organizations are now looking for competencies in their employees with versatile skills, able to comprehend complex internal and external relationships, and the ability to adapt to effective teamwork.

Based on the principle of the peaceful use of atomic energy, the Turkish Nuclear Regulatory Authority (NDK) determines the principles to be applied and the responsibilities of the parties for the protection of employees, the public, the environment and future generations from the possible harmful effects of ionizing radiation during the execution of activities related to nuclear energy and ionizing radiation.

As a result of high-level breakthroughs in the field of nuclear energy in recent years, regulatory and supervisory activities were reorganized within the NDK by taking into account international requirements.

The activities, subjects and fields to be regulated by NDK are listed below:

- Radiation protection of workers, the public, the environment and future generations;
- Safety, security and safeguards in activities related to nuclear energy and radiation;
- All activities related to the establishment, operation, decommissioning or closure of a nuclear facility, radiation facility or radioactive waste facility;
- Extraction, production, transportation, storage, export, import, trade, possession, transfer, processing, reprocessing and use of nuclear material;
- Production, transportation, storage, export, import, trade, possession, transfer, use, installation, modification, disassembly, maintenance and repair of radiation sources;
- Possession, transfer, processing, transportation, storage, export, import and disposal of radioactive waste;
- Export and import of substances, materials, equipment, systems, components or related technology determined by NDK to be within the scope of safeguards;
- Radiation emergency management;
- Qualification and training of personnel related to activities falling within the scope of their duties and authority;
- Other subjects, areas and activities that fall under the NDK's mandate and jurisdiction, to be determined by the Board of NDK.

### 5.2.1. Organization and human resources

The NDK is composed of a Nuclear Regulatory Board and the Presidency. The decision-making organ of the NDK is the Nuclear Regulatory Board consisting of five members, including the President of NDK (who also chairs the Board) and a Second Chairperson. The President of the NDK, the Second Chairperson and all Board members are assigned by the President of the Republic. The Regulation on Working Procedures and Principles of the Nuclear Regulatory Board was issued in 2022.

The Presidency consists of the President of the NDK, two vice presidents, and multiple units (five technical and six administrative).

The technical units are as follows:

- Department of Nuclear Installations (regulatory activities in nuclear safety);
- Department of Radiation Applications (regulatory activities in radiation applications and radiation facilities);
- Department of Security and Safeguards (regulatory activities in nuclear security and safeguards and transportation and import/export of radioactive materials);

— Department of Radiation Protection (regulatory activities in radiation protection and radioactive waste management safety);
— Department of Inspection (nuclear and radiation safety inspections).

The administrative units are as follows:

— Department of External Relations (national and international coordination of all kinds of activities within the scope of duties and responsibilities of NDK);
— Department of Legal Services;
— Department of Strategy Development (administrative and financial activities of NDK);
— Department of Support Services (human resources and other supporting services);
— Press and Public Relations Consultancy;
— Board Services Director.

Additionally, the Directorate of Board Services conducts secretarial work and operations of the Nuclear Regulatory Board.

### 5.2.2. Capacity building training programme

*5.2.2.1. Basis for training programme*

The objective of the NDK's training programme is that the staff of the regulatory body develop their administrative and technical qualifications and competencies within the scope of a training management system. Effective training programmes have a direct impact on corporate success and country development as well as developing successful employees. Long term training plans are made, taking into account the needs of the regulatory body. They are prepared in accordance with the personnel policy and approved by the Board of the NDK.

The regulation of training in public institutions and organizations in Türkiye is obliged by law (Civil Servants Law No. 657 adopted in 1965). The seventh part of the law regulates the principles of raising the civil servants of institutions in service under the title of 'training of civil servants' of Articles 214 to 225.

The principles to be applied to achieve the goals specified in the training are as follows:

— Programmes related to educational activities, within the framework of the Development Plan, Institution Strategic Plan, Civil Servants' Training Plan and the training plan to be prepared in line with this programme and in line with the objectives, duties and responsibilities of NDK, training needs analysis, surveys and evaluations, and the duties and seniority of personnel;
— Conduct of the training in accordance with the principles of effectiveness, frugality and efficiency, being renewable and continuous;
— Cooperation with all relevant organizations in education when necessary, ensuring the exchange of information, documents, tools and trainers;
— Organization of the annual training plan, training programme and training activities in a way that does not disrupt the main duties of the NDK;
— Participation of all personnel in training activities according to the determined training needs, based on priorities in benefiting from training and taking into account the principle of equality of opportunity;
— Responsibility of staff for participating in and continuing the training programme;
— Suitability of the environments for the training needs and availability of the necessary equipment;
— Provision of on-the-job practical training in order to reinforce the knowledge and skills gained in education, when necessary;
— Surveys, observation, interviews and evaluations during and after the training in order to determine whether the results expected from the training have been achieved.

The NDK attaches importance to training activities in order to continuously develop human resources within the framework of following and interpreting developments in an up-to-date way; developing regulations on the basis of innovative, analytical and data-based safety and security; contributing results that are oriented to support decision making processes; and being proactive, with common sense, in order to promote teamwork, learning and specialization. This programme has been prepared for this purpose.

#### 5.2.2.2. *Target audience of training*

The training programme serves the purpose of informing NDK employees about the behaviours expected of them, supporting the personal development of the personnel of the institution, knowing and adopting the quality standards of the system by all employees, standardizing the working methods, increasing the mutual trust of employees and the institution, attracting qualified personnel to the institution, preparing employees for higher level duties and, most importantly, creating a learning culture in the institution. Training is grouped under four headings:

— Training for new employees;
— Vocational training;
— Training for technical staff;
— Activities for the training of managers.

An annual training programme is prepared in accordance with the training plan. Training can take place in the NDK or in other organizations in Türkiye or abroad, as appropriate. The annual training programme is prepared by the NDK Department of Support Services, with input from the other departments, and is approved by the NDK President.

#### 5.2.2.3. *Types of training provided*

Training is given to personnel in order to develop skills related to their duties, increase their practical and theoretical knowledge, train them in service, and ensure efficiency, effectiveness and high quality.

**Education in Türkiye**

Domestic theoretical and practical training opportunities, including graduate education and on-the-job training, are provided to personnel in order to improve their professional knowledge. Educational priorities and needs are determined by the duties and functions of staff at the NDK.

**Study abroad**

Personnel are provided with theoretical and practical training abroad, including on-the-job training, in order to improve their professional knowledge within the framework of the relevant regulation.

**Review and research studies**

Personnel can be sent abroad for on-the-job trainings on review and research with the approval of the relevant units. The duration of the assignment is determined by the NDK President, not to exceed six months according to the scope of the study. The individual receives a detailed programme of work for this period, and prepares a report on the work carried out, which is submitted to NDK within one month of return.

#### 5.2.2.4. *Training courses provided*

Training courses enable personnel to gain the necessary knowledge, skills, attitudes and behaviours related to their duties. Training can help employees to adapt to the needs of the profession, and to progress and develop in the profession.

Training courses provided to the staff commensurate with their role in the organization are defined as the following:

- Training for new employees:
  - Pre-service training;
  - Adaptation and orientation training;
  - On-the-job training.

- Vocation training:
  - Gaining professional skills;
  - Basic vocational training.

- Training for technical staff:
  - Technology adaptation training;
  - Progress and development in the profession;
  - Advanced vocational training.

- Activities for the training of managers:
  - Management sciences and technology education;
  - Training on aspects such as research and development, problem solving;
  - Technology, evaluation and adaptation training;
  - Quality management training;
  - Change management training;
  - Crisis management training;
  - Organization development training;
  - Management training by purposes;
  - Time management training;
  - Stress management training;
  - Conflict management training;
  - Project management training.

Additionally, a training programme for the site evaluation team is carried out within the scope of the team training programme for the safety assessment of a site for nuclear installations.

The training programme also helps the understanding of EIA, emergency planning considerations and management of siting activities. The training programme does not contain any specific training for siting inspection personnel. Inspector training, also including siting inspections, are discussed within the scope of the specific training for candidate inspectors.

For members of the safety assessment team for the sites of nuclear installations, competences in the following areas are beneficial:

- Demography and population distribution;
- Meteorology;
- Hydrology and hydrogeology;
- Environmental monitoring and EIA;
- Geology;
- Seismology;
- Volcanology;
- Geotechnics, earthwork and foundation engineering;
- External human induced event assessment;
- Security;
- Analysis of feasibility of emergency planning;
- Land use;

— Grid infrastructure;
— Oceanography.

Additionally, competence in the following disciplines helps to support the safety assessment team for the sites of nuclear installations:

— Human resources management;
— Stakeholder involvement;
— Procurement;
— Legal issues;
— Project management;
— Quality management;
— Authorization and licensing;
— Environmental law.

*5.2.2.5. Frequency and duration of training courses*

The frequency and duration of the training courses are tailored to the specific needs at the time. It may be performed online or in person. The duration of the training programme may be a single day or may continue up to a duration of months or years. Relevant examples are provided in Annex II of this publication.

*5.2.2.6. Evaluation of success of training*

In order to ensure that the training is successful after its completion, questionnaires are completed. The topics addressed in the questionnaires include whether the training objectives were met, the satisfaction of the participants, the quality of the training and suggestions for the improvement of future training.

*5.2.2.7. Short term and long term (continuing) training*

The main purpose of the training is to ensure that employees take an active role in future work by placing a culture of continuous learning and development in the NDK. The need to increase the service quality of the institution and to develop continuously has turned into an organized training activity in accordance with needs. Rapid information and technology developments also have a great impact on training. Knowledge and skills transfer based on mentoring has been supported by educational activities using all kinds of audiovisual tools and distance learning methods in educational technology.

*5.2.2.8. Mechanism for feedback and continuous improvement*

The result of the evaluation survey mentioned in subsection 5.2.2.6 is incorporated into future training programmes and they are prepared in line with the lessons-learned to support the regulatory activities.

*5.2.2.9. Follow-up training*

Training programmes are arranged considering the seniority, planned duties and responsibilities of the personnel. Future trainings are performed on a regular basis in line with the needs analysis.

### 5.2.3. Knowledge management

The NDK established its own knowledge management system in order to ensure that its assigned responsibilities are properly fulfilled, to maintain and improve its performance, to foster and support a safety culture, and to share knowledge and experience within the organization, also making use of the mentoring approach and relevant training programmes.

The knowledge management system uses an organizational manual, procedures and guides regarding all activities within NDK and management review of the system with the aim of continuous improvement of its operation.

## 5.3. UNITED ARAB EMIRATES – FEDERAL AUTHORITY FOR NUCLEAR REGULATION

The Federal Authority for Nuclear Regulation (FANR) was established in 2009 with the goal of contributing to the future growth and prosperity of the United Arab Emirates (UAE) and its people by meeting the highest standards of nuclear safety, security and safeguards. Since its inception, FANR has successfully fulfilled its mandate to ensure the safe, secure and peaceful use of nuclear energy and radiation sources in the UAE.

FANR remains dedicated to developing Emiratis in the nuclear sector, and this forms part of its capacity building and sustainability efforts.

### 5.3.1. Organization and human resources

FANR has made significant progress in recruiting a qualified and capable workforce. FANR is committed to optimizing the skills, processes and resources needed to excel and realize the authority's vision. FANR capacity-building efforts include steadfast support of the Government's 'Emiratisation' initiative. The main goal of all education and training activities at FANR is to ensure Emiratis are trained and developed to have the necessary skills and knowledge to contribute effectively to FANR's core functions.

FANR's organizational structure includes two divisions: Administration Division and Operations Division. The Administration Division includes the departments of Finance and Control, Human Resources, Information and Communication Technology, Government Communications, and Supply Chain and General Services. The Operations Division includes the departments of education and training, Nuclear Safety, Nuclear Security, Radiation Safety, and Safeguards.

Long-term career opportunities for Emirati employees at FANR are encouraged through focused recruitment, knowledge transfer and training and development programmes.

As of the date of this report, there are 245 FANR staff members. The balance of the staff includes expatriates with nuclear experience recruited from 28 countries around the world. The depth and breadth of expertise embodied within this team has been instrumental to FANR's achievements to date.

As of the date of this report, the number of qualified inspectors at FANR is 88. They are involved in regulatory activities, supported by other FANR subject matter experts and contractors. FANR has now deployed 8 resident inspectors to a permanent site office at Barakah NPP to assist in overseeing the construction, commissioning and operation activities taking place at the site.

The siting section is part of the nuclear safety department, and intensive interaction at early stage during construction and design group from nuclear safety department is highly recommended.

### 5.3.2. Capacity building training programme

#### *5.3.2.1.Measures to develop and maintain competence*

Having successfully recruited a workforce to meet near-term demands, FANR's human resource strategy for long-term sustainability concentrates on developing Emiratis to take positions of increasing responsibility while retaining an appropriate cadre of non-Emirati experts. FANR complements its in-house training programmes through collaboration with Emirates Nuclear Energy Corporation, Khalifa University, the IAEA and other partner institutions in a national programme of capacity building, which offers Emiratis a range of education, training and development opportunities in the UAE and abroad.

#### *5.3.2.2.Internship*

In order to enhance FANR's capacity building and staff competence, FANR awards scholarships to Emirati employees to complete tertiary qualifications at leading institutions such as Zayed University, Manchester University, KINS, the Korea Advanced Institute of Science and Technology and Khalifa University.

The Employee Development Programme is designed to support the development of FANR's employees by equipping them with the knowledge and skills needed to perform their roles and responsibilities.

#### *5.3.2.3. Trainee Engineers Programme*

FANR has pursued the implementation of its programme for fresh graduates with its 'Trainee Engineers Programme'. This programme is designed to provide fresh Emirati engineering and physics graduates with the fundamental knowledge, skills and attitudes necessary to understand technical concepts applicable to nuclear engineering, radiation safety and, specifically, nuclear regulation. This development programme runs over 53 weeks and is comprised of intensive training on nuclear and radiation fundamentals, on-the-job training in each of the departments in the Operations Division as well as soft skills training. Upon completion of the programme, Emirati nuclear engineers and physicists will be fully integrated into the departments in the Operations Division. Over the last few years, six to twelve young engineers per year have benefited from this programme. FANR's employee development programme consists of six serial stages.

**(1) Onboarding programme (2 weeks)**

This programme introduces trainees into FANR business including an introduction on each department. The topics to be covered in the onboarding programme consists of an induction into FANR business (2 days), fundamentals of nuclear law/regulatory framework (2 days), safety and security culture (1 day), introduction to nuclear (4 days, and Barakah NPP visit (1 day).

**(2) Technical training (6-12 weeks)**

This programme provides detailed technical lectures and practices. Training topics included in this programme are radiation safety training and nuclear energy management school (see Annex II for more detailed information about specific training courses).

**(3) Job rotations (32 weeks)**

The job rotation plan provided by the rotation department includes a project/assignment. A mentor from each department is identified by the Director of that department to provide on-the-job mentorship to the trainees. Trainees are expected to provide a summary report of what they learnt during this rotational assignment.

**(4) Coaching and mentoring (ongoing)**

Each trainee receives a technical mentor from each department and an education and training coach through the programme. This is to ensure the trainees stay on track with their development plan.

**(5) Behavioural soft skills training (2 weeks)**

To ensure the trainees are well-rounded and ready for the workplace, they are assigned to complete various behavioural skills training courses. This training includes effective communication, emotional intelligence, and teamwork and collaboration.

**(6) Study tour (1-2 weeks)**

The study tour marks the start of the external training abroad. Trainees get to experience an actual plant walkdown, visit various radiation facilities, spend time with the nuclear and radiation regulator and engage with subject matter experts at the facilities.

*5.3.2.4. Competency development framework*

The competency development framework is the foundation for continuously ensuring a highly competent workforce at FANR. In 2021, technical and behavioural competency assessments for both divisions were conducted in an effort to highlight focused training and development needs. Following the 2020 launch of the competency framework automation system (where employees and their managers conduct online competency assessments) the competency framework was also integrated into the human resources performance appraisal system. This provided an additional monitoring tool to ensure continuous development is maintained through performance objectives on an annual basis. As of 2021 the Competency Framework Automation System ensures that all training and development activities are linked with the gap analysis and assessment.

### 5.3.3. Knowledge management

FANR has established a knowledge management programme for business sustainability purposes. This programme has been designed to support its knowledge-based decisions for the safe and efficient regulation of nuclear and radiation safety activities in the UAE.

The knowledge management programme also aims to support FANR management by minimizing the impact of employee mobility (e.g. transfer of personnel within or outside the organization, retirement, promotion) and the resulting knowledge loss. It also aids the transfer of nuclear knowledge from one employee to another. In order to achieve these objectives, FANR has developed a number of methodologies and tools during the past two years.

For example, FANR has developed tools like the Knowledge Loss Risk Assessment and the Knowledge Transfer Plan (KTP) which support identification of employees with critical knowledge and accordingly transfer the targeted knowledge to their successors. The use of these methods and tools will help in mitigating the related business risks and support business continuity. Moreover, based on the lessons learned over the years, the Knowledge Loss Risk Assessment tool has evolved and has been automated and integrated with the competency framework which was developed based on Ref. [13].

## 5.4. UNITED STATES OF AMERICA – NUCLEAR REGULATORY COMMISSION

The United States Nuclear Regulatory Commission (NRC) regulates commercial NPPs and other uses of nuclear material, such as in nuclear medicine, through licensing, inspection and enforcement of its requirements. The NRC is headed by five Commissioners appointed by the President and confirmed by the Senate for five-year terms. The NRC formulates policies, develops regulations governing nuclear reactor and nuclear material safety, issues orders to licensees, and adjudicates legal matters.

The NRC's commercial reactor licensing and inspection activities include the following:

- Reviewing applications for proposed new reactors;
- Inspecting construction activities;
- Reviewing separate licence change requests, called 'amendments', from power reactor licensees;
- Performing inspections at each operating reactor site;
- Conducting initial reactor operator licensing examinations;
- Ensuring NRC-licensed reactor operators maintain as current their knowledge and skills by passing rigorous requalification exams every 2 years and obtaining an NRC licence renewal every 6 years;

— Reviewing operating experience items each year and sharing lessons learned that could help licensed facilities operate more effectively;
— Issuing notices of violation, civil penalties, or orders to operating reactors for significant violations of NRC safety and security regulations.

There are four regional offices in Pennsylvania, Georgia, Illinois and Texas that oversee day to day operation of the major nuclear facilities that the agency regulates. The regional offices support at least two resident inspectors at each NPP site.

The NRC establishes requirements for the design, construction, operation and security of US commercial NPPs. It ensures that plants operate safely and securely within these requirements by licensing the plants to operate, licensing control room personnel, establishing technical specifications for each operating plant, and inspecting plants daily.

### 5.4.1. Organization and human resources

There are several offices and committees reporting to the NRC. Some of the main organizations include:

— Office of Congressional Affairs: responsible for providing information related to legislative matters affecting the NRC in the US Congress to the NRC and all its employees.
— Office of the Chief Financial Officer: responsible for planning, appropriating use of financial resources and providing financial services to support the NRC staff.
— Office of the General Counsel: responsible for legal advice and counsel to the NRC and its staff.
— Advisory Committee on Reactor Safeguards: an independent body required by US law to provide the NRC with independent technical advice on issues of public safety related to nuclear reactors, reactor safeguards, and nuclear waste and materials management issues.
— Executive Director for Operations: the chief operating officer of the Commission, responsible for supervision and coordination of policy development, agency operational activities, and implementation of NRC policy directives.

The NRC's major programme components are the Office of Nuclear Reactor Regulation, the Office of Nuclear Material Safety and Safeguards, and the Office of Nuclear Regulatory Research.

The Office of Nuclear Reactor Regulation is responsible for the NRC's nuclear reactor safety mission. The Office's regulatory activities include licensing, oversight, siting, rulemaking, and incident response for operating commercial nuclear power reactors, new commercial nuclear power reactors, advanced reactor technologies, and non-power production and utilization facilities. The Office of Nuclear Regulation collaborates with other staff in headquarters and regional offices to accomplish the NRC's mission.

The human resources needed for the licensing process of a new NPP are a function of the overall activities performed by the regulatory staff, which include reviewing applications that propose the location of a new plant to ensuring the final safe operation of the facility.

**Overview of the organization and regulatory activities related to proposed nuclear power plants at the United States Nuclear Regulatory Commission**

An overview of the NRC regulatory activities associated with the licensing process for a new NPP is provided in this section.

The application for construction of a new NPP includes the option of an early site permit for either the one-step or two-step application process described below. An early site permit resolves the issues related with site safety, environmental protection, and emergency preparedness issues independent of a specific design for a nuclear installation. The early site permit application has to address the safety and environmental characteristics of the site and evaluate potential obstacles to developing an acceptable emergency plan. The application covers the following information:

— Site boundaries;
— Seismic, meteorological, hydraulic and geological data;
— Existing and projected future population of the surrounding area;
— Evaluation of alternative sites;
— Proposed general location of each plant planned to be on the site;
— Number, type and power level of the plants planned for the site;
— Maximum discharges from the plant;
— Type of plant cooling system to be used;
— Radiation dose consequences of hypothetical accidents;
— Plans for coping with emergencies.

The NRC has two separate processes for licensing a new power reactor: a combined licence process in accordance with 10 CFR Part 52 [32], and a two-step process in accordance with 10 CFR Part 50 [33]. The requirements for both processes are similar.

**Combined licence**

A combined licence under 10 CFR Part 52 [32] authorizes construction of the facility and contains essentially the same information required in an application for an operating licence issued under Part 50 [33]. It specifies the inspections, tests and analyses that the applicant has to perform. It also specifies acceptance criteria necessary to provide reasonable assurance that the facility has been constructed and will be operated in agreement with the licence and applicable regulations. An early site permit or design certification can also be reviewed by the NRC during this process. The NRC reviews the technical and environmental information as described for the two-step licensing process.

After issuing a combined licence, the NRC authorizes operation of the facility only after verifying that the licensee has completed required inspections, tests and analyses and that acceptance criteria have been met.

**Two-Step licensing process**

In order to use the two-step process, an application for construction or operation of an NPP requires a SAR. This document contains the design information and criteria for the proposed reactor, and comprehensive data on the proposed site. It also discusses various hypothetical accident situations and the safety features of the plant that would prevent accidents or lessen their effects. In addition, the application has to contain a comprehensive assessment of the environmental impact of the proposed plant.

The NRC staff reviews the application to determine whether the plant design meets all applicable regulations (10 CFR Parts 20 [34], 50 [33], 73 [35], and 100 [36]). The review includes site characteristics such as surrounding population, seismology, meteorology, geology and hydrology. The review also includes:

— Design of the NPP;
— The plant's anticipated response to hypothetical accidents;
— Plant operations, including the applicant's technical qualifications to operate the plant;
— Discharges from the NPP into the environment (i.e. radiological effluents);
— Emergency plans.

### 5.4.2. Capacity building training programme

In view of the diverse needs of the NRC staff, there are generic training and qualification requirements followed by job specific training related to the tasks to be performed by the personnel. The NRC

Technical Training Center develops and implements the policy and programme for training of NRC inspectors and provides technical training for NRC personnel and other government agencies as requested. The Technical Training Center provides or coordinates training in NPP design and operation; inspection techniques; physical security; health physics; and such disciplines as welding, non-destructive testing, and quality assurance. Training in a control room simulator is an important part of the NRC's training programme. The Technical Training Center has several light water reactor control room simulators to provide training of inspectors, operator licence examiners and other technical personnel in transient and accident response, extended scenarios including prolonged operation in emergency operating procedures, analyses for postulated scenarios, and support of various human factors and research projects.

The NRC has formal training and development programmes which provide a means of defining, capturing and disseminating the information and knowledge that is needed to carry out the duties for a specific position or activity. They are a mechanism for ensuring that relevant organizational knowledge, experiences and lessons learned are transferred to replacement staff in a reliable and systematic fashion.

The well-established programmes include formal training, qualification, and development programmes, including the following:

- New Employee Orientation;
- New Employee Orientation;
- Nuclear Safety Professional Development Programme;
- Honour Law Graduate Programme;
  Leadership Potential Programme;
- Team Leader Development Programme;
- Senior Executive Service Candidate Development Programme;
- Acquisition Project Manager Certification;
- Inspector Qualification;
- Instructor Qualification;
- Headquarters Operations Officer Qualification;
- Headquarters Emergency Response Officer Qualification.

In addition to classroom training activities, many of these programmes include on-the-job training elements that facilitate tacit knowledge transfer from more experienced staff to newer staff.

The generic qualification requirements for technical staff consist of study activities on a variety of subjects including technical, regulatory, administrative, data searches, public interface and enforcement. These study activities are designed to focus on necessary regulatory knowledge, skills and abilities. Evaluation criteria have been established in each study activity to indicate that pertinent information has been learned from the references and documents in the training programme.

Technical employees of all levels and experience who are working to complete any of the required qualification programmes have administrative procedures that define the training and qualification needed to enable the staff to perform regulatory activities. Various disciplines develop specific training materials on unique aspects necessary in their areas of expertise.

Resident inspectors and staff responsible for routine inspections have specific training programmes to familiarize them with the regulatory provisions, plant licensing and design basis necessary for licensees to operate their facilities in a safe manner. The training aids the inspectors to master the techniques and skills needed to collect, analyse and integrate information, understand the technology, apply concepts in various technical areas, and develop and maintain an understanding of how basic NPP design and operations provide for protection of public health and safety.

The NRC has arrangements with equipment manufacturers (e.g. of valves, pumps, emergency diesel generators, electrical protective equipment and switchgears) to provide component specific training on equipment important to safety. This training includes manufacturer recommended maintenance and surveillance activities and performance capabilities.

#### *5.4.2.1. Nuclear power plant operator training*

The NRC is the oversight body for the commercial nuclear industry. 10 CFR Part 55 [37] contains the requirements for licensees to operate nuclear facilities. The regulation requires NPP operators to go through a rigorous training programme before being licensed to operate equipment in the NPP.

The training process involves gaining a broad and deep knowledge of the reactor systems and how they affect each other, and operation of the reactor. Trainees spend several months reading about the reactor systems and drawing components to help them to understand how everything is connected.

When the trainee has gained sufficient knowledge of the reactor, they start to operate the reactor under direct supervision from the training supervisor. Trainees learn to operate the reactor by performing start-ups, shutdowns, power manipulations and control blade shuffles. Much like driving a vehicle, learning how quickly the reactor responds to control blade movement and temperature change can only be gained by practice.

The training programme involves classroom training on power plant systems including nuclear science fundamentals. A simulator is used to train on procedures used for reactor startup and verification of reactor response. Classroom and simulator training is used to understand abnormal and emergency operating procedures followed by approximately 600 hours of on-shift training under the supervision of licensed operators. The training programme includes:

- Systematic analysis of the jobs to be performed;
- Learning objectives derived from analysis, which describe desired performance after training;
- Training design and implementation based on the learning objectives;
- Evaluation of trainee mastery of the objectives during training;
- Evaluation and revision of the training based on the performance of trained personnel in the job setting.

NRC tests include questions on system knowledge and operation, and abnormal and emergency operating procedures. The simulator examination includes observation of crews of reactor operators demonstrating capability to safely shut down a plant following a severe accident and other abnormal conditions.

Once the trainee has demonstrated the ability to operate the reactor safely, including responding appropriately to abnormal conditions, the trainee applies for their licence. The licence exam is conducted over two days by an NRC examiner and includes a multiple-choice written exam, an interview, a tour of the reactor and a reactor startup. If the trainee passes the exam, then they become a licensed reactor operator.

### 5.4.3. Knowledge management

Knowledge management is an integral part of NRC's culture. Employees capture and share knowledge and experiences as part of routine job functions and mentor and coach peers via formal or informal training sessions. The NRC has resources and tools in place that help an individual or organization to determine the optimum methods for implementing knowledge management practices and ensure the NRC continues to build, retain and sustain the staff's organizational skills.

The knowledge management programme at NRC uses a broad and continuously evolving range of methodologies for identifying, collecting and sharing knowledge. The methods include stored technical documents, job aids and desk references, written policies and procedures, regulatory guides, standard review plans, and regulatory issue summaries. The strategies used to facilitate knowledge transfer include training courses, formal and informal mentoring, rotational assignments, internal seminars and presentations, invitational seminars and panel discussions, video interviews, newsletters and websites. Figure 4 presents the knowledge management programme.

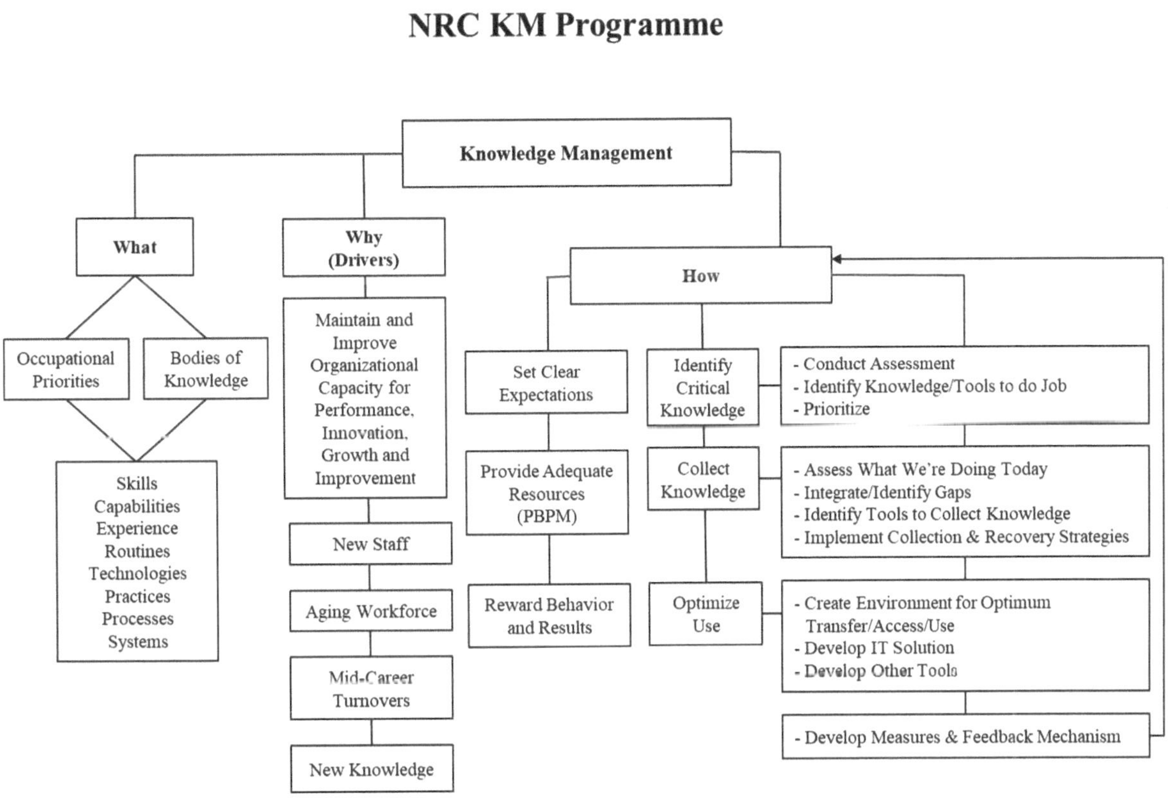


*FIG. 4. The NRC knowledge management programme [38]. KM: knowledge management; PBPM: planning, budgeting, and performance management.*

The Technical Training Center and Professional Development Center sponsor numerous courses designed to maintain the important skills and knowledge of NRC staff. These courses are developed using both NRC and external subject matter experts to capture unique knowledge and experience and to disseminate that knowledge widely throughout the NRC. Individual offices and regions also provide locally developed and administered training, operational review sessions and seminars for their staff on specific topics that aid in transferring knowledge from experienced to less experienced employees and as a method of transferring lessons learned from events to the staff. Under the knowledge management programme the NRC staff provide knowledge-sharing seminars on technical topics both Commission-wide, for subjects that have a broad application or interest, and on a smaller scale for technical topics that primarily relate to the work of an individual organization.

The NRC sometimes meets short-term needs to fill current knowledge gaps by contracting or hiring expert consultants to provide the needed expertise to accomplish specific tasks. Contractors and expert consultants are usually employed where specialized knowledge and skills are needed by the NRC but cannot be obtained through normal recruitment channels. The contractors or consultants fill current knowledge gaps and may also focus on transferring their knowledge to NRC staff before the end of the contract.

The NRC has developed e-learning as part of its remote learning tools, and it has proven to be very productive. Though it is not a universal replacement for traditional classroom instruction it is an effective and economic way to deliver just-in-time, focused learning solutions. Under the knowledge management programme the NRC staff identify and separate the training objectives that can be delivered through e-learning applications from those objectives that are more effectively delivered through more traditional methods.

To make e-learning became an important tool for transferring knowledge, the NRC upgraded video teleconferencing and other meeting support technologies to enhance meeting capabilities and reduce travel time and expenses. As part of knowledge transfer, former and current NRC experts are invited to participate in seminars and panel discussions. They share stories of their past experience on various relevant topics to give a historical perspective to NRC's policies, processes and procedures, and significant past events. The sessions are video recorded and made available as a knowledge resource on the NRC's intranet.

The NRC has implemented cloud-based data-centre capabilities as a means to consolidate services and offer knowledge management courses to all staff. An example of such a tool is an online encyclopaedia called 'Nuclepedia' that provides a collaborative forum for employees to capture and share information. Nuclepedia is a platform that facilitates NRC experts to provide technical position papers describing specific reasons and basis for aspects of nuclear safety and lessons learned. In addition to capturing valuable information from experienced employees, Nuclepedia is also a platform that is searchable, easy to contribute to, and accessible to all NRC staff.

The NRC developed the NUREG/KM series in 2012, which is a series of publications established to preserve knowledge of documents and events that shaped the regulatory process or a technical topic. The series focuses on collecting and interpreting historical information on identified topics for the benefit of future generations of NRC professionals as well as the public. Currently, the series features four publications: (1) NUREG/KM-0001, "Three Mile Island Accident of 1979 Knowledge Management Digest-Overview" [39], (2) NUREG/KM-0002, "The Browns Ferry Nuclear Plant Fire of 1975 Knowledge Management Digest" [40], (3) NUREG/KM-0004, "Fuel Behavior under Abnormal Conditions" [41], and (4) NUREG/KM-0006, "Fundamental Theory of Scientific Computer Simulation Review" [42].

# 6. SUMMARY

## 6.1. CAPACITY BUILDING PROGRAMME

The infrastructure and processes for education and training are fundamental to the capacity building strategy of Member States, as they provide a structure to develop the capacity of those individuals involved in the application of nuclear technologies. This involves an integrated effort between the technical organization, human resources management, and knowledge management. Continuous improvement is a necessary component of an effective training programme; therefore, it needs to be incorporated into the management system of the organization and subject to quality control, record keeping, feedback and other aspects of the management system.

Because of the natural turnover of the organization, as well as the special challenge of the early labour force of nuclear safety professionals retiring out of active service, regulatory bodies (and other organizations) responsible for nuclear safety need to strengthen, develop, maintain and implement their capacity building programmes.

A structured process to develop a capacity building programme consists of the following components:

— Assessment of training needs – determined by a review and analysis of the responsibilities of the organization, as well as consideration of where and when there may be a potential shortfall of trained personnel. At a minimum, training of new personnel and refresher training of existing personnel are necessary;
— Design of training programme – the objectives of the training programme are clearly delineated in designing the training programme to assure that nuclear safety can be assured within the

responsibilities of the organization being trained. Training modules are prioritized to fit within the overall goals and time frame of the organization;

- Development and implementation of the training programme – for the training of existing personnel, adequate provisions need to be made to cover the responsibilities of the individual while training is being undertaken. Additionally, on-the-job training impacts the location of the training as well as all involved. Consideration is also given to the development of effective training materials. An implementation schedule is arranged that meets the needs of the capacity building programme as well as the constraints of the operating organization;
- Evaluation of effectiveness of training programme – objective metrics are developed to determine the effectiveness of training programme, such as the time for trainees to obtain appropriate certifications, reduced errors after receiving training, and time needed to be spent in training to achieve the desired results. Additionally, feedback from trainers, trainees and their supervisors all help in evaluating training programmes and ensuring their continuous improvement.

## 6.2. SPECIFIC TRAINING ON SITE EVALUATION FOR NUCLEAR FACILITIES AND ACTIVITIES

Most organizations offer general training on nuclear safety along with specific training on site evaluation. General training is provided to all staff, at a frequency and duration appropriate to the individual's job responsibilities vis-à-vis nuclear safety. Specific training on site evaluation is provided in accordance with an individual's job responsibilities. Training may be in house, provided by external experts, or completed through university training, perhaps even external to the Member State. Some universities work with nuclear programme providers to establish a course structure that covers the many skills needed to enter the nuclear industry. Many options exist and may be appropriate based on the needs of the organization and the gaps that have been identified in the training programme development.

For an embarking country, training may be largely external to the organization; whereas for a country with an established nuclear power programme, much of the training may be provided internally within the organization, by in-house trainers.

Capacity building is a complex undertaking that needs continuous attention, understanding of the many responsibilities of the organization regarding nuclear safety, knowledge of the workforce and a solid analysis of the gaps in the organization's training needs and availability of trained personnel. Therefore, separate divisions focused on capacity building are generally created within the organization, associated with the human resources division or the operating division, to assure that the workforce is adequately trained to meet the organization's responsibilities for nuclear safety.

## APPENDIX

Table 3 provides a list of areas that might be included in an education and training programme on the review of nuclear installation site safety.

TABLE 3. LIST OF AREAS FOR POSSIBLE INCLUSION IN A SITE EVALUATION TRAINING PROGRAMME

| Area | Course contents | IAEA reference documents |
|---|---|---|
| Fundamental safety principles | Safety objectives<br>Safety principles | SF-1 [4] |
| Legal and regulatory framework | Legal framework<br>Regulatory framework<br>Roles of regulatory body<br>Roles of operating organization | GSR Part 1 (Rev. 1) [6] |
| Leadership and management for safety | Leadership<br>Management<br>Safety culture | GSR Part 2 [7] |
| Licensing process | Licensing process<br>Role of regulatory body<br>Role of licensee<br>Licensing process for siting<br>- Site survey<br>- Site selection | GSR Part 1 (Rev. 1) [6], and SSG-12 [11] |
| Review and assessment | Review and assessment<br>- Process, procedure, guidelines<br>- Documents to review<br>Site evaluation<br>- Site survey<br>- Site selection | GSR Part 1 (Rev. 1) [6], SSG-12 [11], SSR-1 [5], SSG-35 [29], and Other relevant publications |
| Inspection | Inspection<br>- Process, procedure, guidelines<br>- Documents to review/inspect<br>Site inspection<br>- Site survey<br>- Site selection | GSR Part 1 (Rev. 1) [6] SSG-12 [11], SSR-1 [5], and SSG-35 [29] |
| Enforcement | Enforcement | GSR Part 1 (Rev. 1) [6], and SSG-12 [11] |
| Site licence | Siting process<br>- Site characterization stage<br>- Pre-operational stage<br>- Operational stage<br>Site safety assessment and EIA<br>- Radiological study (incl. baseline study)<br>- Important factors<br>Monitoring and periodic review of the site<br>Application of management system for site evaluation | GSR Part 1 (Rev. 1) [6], SSG-12 [11], SSR-1 [5], SSG-35 [29], and GSR Part 2 [7] |
| Other areas (e.g. radiation safety, safety assessment) | Radiation protection<br>Safety assessment<br>Other relevant topics | GSR Part 3 [43], and GSR Part 4 (Rev. 1) [44] |

## REFERENCES


[1] IAEA Action Plan on Nuclear Safety, IAEA, Vienna (2011), https://www.iaea.org/sites/default/files/actionplanns.pdf

[2] Note by the Secretariat, Strategic Approach to Education and Training in Nuclear Safety 2013–2020, 2013/Note 9, IAEA, Vienna (2013).

[3] INTERNATIONAL ATOMIC ENERGY AGENCY, IAEA Report on Capacity Building for Nuclear Safety, Action Plan on Nuclear Safety Series, IAEA, Vienna (2015).

[4] EUROPEAN ATOMIC ENERGY COMMUNITY, FOOD AND AGRICULTURE ORGANIZATION OF THE UNITED NATIONS, INTERNATIONAL ATOMIC ENERGY AGENCY, INTERNATIONAL LABOUR ORGANIZATION, INTERNATIONAL MARITIME ORGANIZATION, OECD NUCLEAR ENERGY AGENCY, PAN AMERICAN HEALTH ORGANIZATION, UNITED NATIONS ENVIRONMENT PROGRAMME, WORLD HEALTH ORGANIZATION, Fundamental Safety Principles, IAEA Safety Standards Series No. SF-1, IAEA, Vienna (2006), https://doi.org/10.61092/iaea.hmxn-vw0a.

[5] INTERNATIONAL ATOMIC ENERGY AGENCY, Site Evaluation for Nuclear Installations, IAEA Safety Standards Series No. SSR-1, IAEA, Vienna (2019).

[6] INTERNATIONAL ATOMIC ENERGY AGENCY, Governmental, Legal and Regulatory Framework for Safety, IAEA Safety Standards Series No. GSR Part 1 (Rev. 1), IAEA, Vienna (2016).

[7] INTERNATIONAL ATOMIC ENERGY AGENCY, Leadership and Management for Safety, IAEA Safety Standards Series No. GSR Part 2, IAEA, Vienna (2016), https://doi.org/10.61092/iaea.cq1k-j5z3.

[8] INTERNATIONAL ATOMIC ENERGY AGENCY, Establishing the Safety Infrastructure for a Nuclear Power Programme, IAEA Safety Standards Series No. SSG-16 (Rev. 1), IAEA, Vienna (2020).

[9] INTERNATIONAL ATOMIC ENERGY AGENCY, Functions and Processes of the Regulatory Body, IAEA Safety Standards Series No. GSG-13, IAEA, Vienna (2018).

[10] INTERNATIONAL ATOMIC ENERGY AGENCY, Organization, Management and Staffing of the Regulatory Body for Safety, IAEA Safety Standards Series No. GSG-12, IAEA, Vienna (2018).

[11] INTERNATIONAL ATOMIC ENERGY AGENCY, Licensing Process for Nuclear Installations, IAEA Safety Standards Series No. SSG-12, IAEA, Vienna (2010).

[12] INTERNATIONAL ATOMIC ENERGY AGENCY, Recruitment, Qualification and Training of Personnel for Nuclear Power Plants, IAEA Safety Standards Series No. SSG-75, IAEA, Vienna (2022).

[13] INTERNATIONAL ATOMIC ENERGY AGENCY, Managing Regulatory Body Competence, Safety Reports Series No. 79, IAEA, Vienna (2014).

[14] INTERNATIONAL ATOMIC ENERGY AGENCY, Guidelines on Devising a Programme for Competence Acquisition and Development among Nuclear Regulators, IAEA-TECDOC-1794, IAEA Vienna (2016).

[15] INTERNATIONAL ATOMIC ENERGY AGENCY, Methodology for Self-assessment of Capacity Building in Member States with Nuclear Power Programmes and Those Planning to Embark on Such a Programme, IAEA, Vienna (2012).

[16] INTERNATIONAL ATOMIC ENERGY AGENCY, Methodology for the Systematic Assessment of the Regulatory Competence Needs (SARCoN) for Regulatory Bodies of Nuclear Installations, IAEA-TECDOC-1757, IAEA, Vienna (2014).

[17] INTERNATIONAL ATOMIC ENERGY AGENCY, Nuclear Power Plant Personnel Training and Its Evaluation, Technical Report Series No. 380, IAEA, Vienna (1996).

[18] INTERNATIONAL ATOMIC ENERGY AGENCY, Systematic Approach to Training for Nuclear Facility Personnel Training: Processes, Methodology and Practices, IAEA Nuclear Energy Series No. NG-T-2.8, IAEA, Vienna (2021).

[19] INTERNATIONAL ATOMIC ENERGY AGENCY, A Methodology to Evaluate the Effectiveness of Training in Nuclear Facilities, IAEA-TECDOC-1893, IAEA, Vienna (2019).

[20] INTERNATIONAL ATOMIC ENERGY AGENCY, Managing Nuclear Safety Knowledge: National Approaches and Experience, Safety Reports Series No. 105, IAEA, Vienna (2021).

[21] CANADIAN NUCLEAR SAFETY COMMISSION, Radiation Safety Training Programs for Workers Involved in Licensed Activities with Nuclear Substances and Radiation Devices, and with Class II Nuclear Facilities and Prescribed Equipment, G-313, CNSC, Ottawa (2006).

[22] UNITED STATES NUCLEAR REGULATORY COMMISSION, Standard Review Plan for Review of License Renewal Applications for Nuclear Power Plants — Final Report, NUREG-1800, Revision 2, USNRC, Washington, DC (2010).

[23] INTERNATIONAL ATOMIC ENERGY AGENCY, Hazards Associated with Human Induced External Events in Site Evaluation for Nuclear Installations, IAEA Safety Standards Series No. SSG-79, IAEA, Vienna (2023).

[24] INTERNATIONAL ATOMIC ENERGY AGENCY, UNITED NATIONS ENVIRONMENT PROGRAMME, Prospective Radiological Environmental Impact Assessment for Facilities and Activities, IAEA Safety Standards Series No. GSG-10, IAEA, Vienna (2018).

[25] INTERNATIONAL ATOMIC ENERGY AGENCY, Geotechnical Aspects of Site Evaluation and Foundations for Nuclear Power Plants, IAEA Safety Standards Series No. NS-G-3.6, IAEA, Vienna (2004).

[26] INTERNATIONAL ATOMIC ENERGY AGENCY, Seismic Hazards in Site Evaluation for Nuclear Installations, IAEA Safety Standards Series No. SSG-9 (Rev. 1), IAEA, Vienna (2022).

[27] INTERNATIONAL ATOMIC ENERGY AGENCY, WORLD METEOROLOGICAL ORGANIZATION, Meteorological and Hydrological Hazards in Site Evaluation for Nuclear Installations, IAEA Safety Standards Series No. SSG-18, IAEA, Vienna (2011).

[28] INTERNATIONAL ATOMIC ENERGY AGENCY, Volcanic Hazards in Site Evaluation for Nuclear Installations, IAEA Safety Standards Series No. SSG-21, IAEA, Vienna (2012).

[29] INTERNATIONAL ATOMIC ENERGY AGENCY, Site Survey and Site Selection for Nuclear Installations, IAEA Safety Standards Series No. SSG-35, IAEA, Vienna (2015).

[30] INTERNATIONAL ATOMIC ENERGY AGENCY, Format and Content of the Safety Analysis Report for Nuclear Power Plants, IAEA Safety Standards Series No. SSG-61, IAEA, Vienna (2021).

[31] INTERNATIONAL ATOMIC ENERGY AGENCY, Dispersion of Radioactive Material in Air and Water and Consideration of Population Distribution in Site Evaluation for Nuclear Power Plants, IAEA Safety Standards Series No. NS-G-3.2, IAEA, Vienna (2002).

[32] UNITED STATES NUCLEAR REGULATORY COMMISSION, Licenses, Certifications, and Approvals for Nuclear Power Plants, 10 CFR Part 52, U.S. Government Printing Office, Washington, D.C.

[33] UNITED STATES NUCLEAR REGULATORY COMMISSION, Domestic Licensing of Production and Utilization Facilities, 10 CFR Part 50, U.S. Government Printing Office, Washington, D.C.

[34] UNITED STATES NUCLEAR REGULATORY COMMISSION, Standards for Protection Against Radiation, 10 CFR Part 20, U.S. Government Printing Office, Washington, D.C.

[35] UNITED STATES NUCLEAR REGULATORY COMMISSION, Physical Protection of Plants and Materials, 10 CFR Part 73, U.S. Government Printing Office, Washington, D.C.

[36] UNITED STATES NUCLEAR REGULATORY COMMISSION, Reactor Site Criteria, 10 CFR Part 100, U.S. Government Printing Office, Washington, D.C.

[37] UNITED STATES NUCLEAR REGULATORY COMMISSION, Operators' Licenses, 10 CFR Part 55, U.S. Government Printing Office, Washington, D.C.

[38] UNITED STATES NUCLEAR REGULATORY COMMISSION, NRC program on knowledge management for liquid-metal-cooled reactors, NUGRE/KM-0007, USNRC, Washington, D.C. (2014).

[39] UNITED STATES NUCLEAR REGULATORY COMMISSION, Three Mile Island Accident of 1979 Knowledge Management Digest – Overview, NUGRE/KM-0001 (Rev. 1), USNRC, Washington, D.C. (2016).

[40] UNITED STATES NUCLEAR REGULATORY COMMISSION, The Browns Ferry Nuclear Plant Fire of 1975 Knowledge Management Digest, NUGRE/KM-0002 (Rev. 1), USNRC, Washington, D.C. (2014).

[41] UNITED STATES NUCLEAR REGULATORY COMMISSION, Fuel Behavior under Abnormal Conditions, NUGRE/KM-0004, USNRC, Washington, D.C. (2013).

[42] UNITED STATES NUCLEAR REGULATORY COMMISSION, Fundamental Theory of Scientific Computer Simulation Review, NUGRE/KM-0006, USNRC, Washington, D.C. (2013).

[43] EUROPEAN COMMISSION, FOOD AND AGRICULTURE ORGANIZATION OF THE UNITED NATIONS, INTERNATIONAL ATOMIC ENERGY AGENCY, INTERNATIONAL LABOUR ORGANIZATION, OECD NUCLEAR ENERGY AGENCY, PAN AMERICAN HEALTH ORGANIZATION, UNITED NATIONS ENVIRONMENT PROGRAMME, WORLD HEALTH ORGANIZATION, Radiation Protection and Safety of Radiation Sources: International Basic Safety Standards, IAEA Safety Standards Series No. GSR Part 3, IAEA, Vienna (2014), https://doi.org/10.61092/iaea.u2pu-60vm

[44] INTERNATIONAL ATOMIC ENERGY AGENCY, Safety Assessment for Facilities and Activities, IAEA Safety Standards Series No. GSR Part 4 (Rev. 1), IAEA, Vienna (2016).

# ANNEX I IAEA TRAINING ON NUCLEAR SAFETY

The IAEA has a variety of training relating to nuclear safety. The training most relevant to this publication are described below.

## I-1. BASIC PROFESSIONAL TRAINING COURSE

The IAEA's Basic Professional Training Course (BPTC) on Nuclear Safety [I-1] was developed in 1998 to provide a broad overview of all the safety concepts and their application to the design and operations of nuclear power plants and research reactors. The BPTC has been used all over the world and part of it has been translated into various languages. The updated 23-module course, launched in 2015, was developed to incorporate all current IAEA safety standards, summarizing their content in plain language and aligning the course with the systematic approach to training. The training material includes textbooks, lecture presentations, case studies and questions to complement the BPTC.

### I-1.1. Target audience

The BPTC is primarily oriented towards junior professionals involved in nuclear safety-related activities. The course also targets highly specialized professionals who lack a broader view of nuclear safety.

### I-1.2. Suggested time frame

The BPTC has been designed to be delivered as either a one-month course (short course) or three-month course (long course). The delivery of the course is adapted to the specific course requirements and time constraints.

### I-1.3. Modules

The BPTC training modules include the following:

— Nuclear physics and nuclear reactor principles;
— Radiation protection in nuclear installations;
— Basic principles of nuclear safety;
— Design of nuclear reactors;
— Safety classification of structures, systems and components;
— Deterministic safety assessment;
— Probabilistic safety assessment;
— Integrated risk informed decision making;
— Siting considerations and environmental impact assessment;
— Operational safety, including operational feedback;
— Operational limits and conditions;
— Plant renewals, modifications and upgrades, ageing;
— Maintenance programme;
— Surveillance programme;
— In-plant accident management;
— Emergency preparedness and response;
— Nuclear fuel cycle;
— Decommissioning;
— Waste management;
— Regulatory control;
— Management system, leadership and safety culture;
— Human performance;
— Public communication.

## I-2. IAEA TRAINING COURSE ON REGULATORY CONTROL OF NUCLEAR POWER PLANTS

The IAEA published a training course in 2002 [I-2], the purpose of which is to provide support in the field of regulatory control of nuclear power plants (NPPs) as well as to support the regulatory bodies of Member States in their own training activities. The target group is the professional staff members of nuclear safety regulatory bodies supervising NPPs and having duties and responsibilities in the following regulatory fields: regulatory framework, regulatory organization, regulatory guidance, licensing and licensing documents, assessment of safety, and regulatory inspection and enforcement. Important topics such as regulatory competence and quality of regulatory work as well as emergency preparedness and public communication are also covered.

Reference [I-2] also presents the key issues of nuclear safety such as defence in depth and safety culture, and explains how these are taken into account in regulatory work, for example, during safety assessments and regulatory inspections. By giving topical examples, Ref. [I-2] also reflects how nuclear safety has evolved over the years on the basis of operating experience feedback and results of safety research. The examples cover development of operating procedures and accident management to cope with complicated incidents and severe accidents to stress the importance of the regulatory role in nuclear safety research.

## I-3. EXAMPLE OF THE OVERALL STRUCTURE AND AGENDA OF AN IAEA TRAINING COURSE ON THE REVIEW OF NUCLEAR INSTALLATION SITE SAFETY

A typical capacity building training course is five days in duration, and consists of the following five modules:

(a) Opening and national presentations on site-specific issues and the status of the national nuclear programme. This provides an opportunity for both the IAEA mission team and the national participants to reach a common understanding of the areas to be emphasized during the training programme.
(b) Lectures on general aspects, presenting IAEA safety standards, concepts, methodologies, examples, case studies, and relevant resources.
(c) Lectures on country-specific technical aspects, identified as having capacity and/or confidence vulnerabilities through the self-assessment for the host organization.
(d) Hands-on practice, during which trainees engage in review activities for the safety for nuclear installations. This includes reviewing a sample SAR in the subchapter related to the technical area they learned about during technical lectures. They will also practice issuing requests for additional information (RAIs), participating in question and answer (Q&A) meetings, and writing a safety evaluation report (SER), which summarizes their safety review results in a team setting.
(e) Feedback to participants and conclusion, presenting evaluations and areas for further improvement for the host organization and/or the Member State, as well as for the training programme.

An example overall structure of an IAEA training course on the review of nuclear installation site safety is given in Table I-1.

TABLE I-1. EXAMPLE OF A TYPICAL OVERALL STRUCTURE OF AN IAEA TRAINING COURSE ON THE REVIEW OF NUCLEAR INSTALLATION SITE SAFETY

| Day | 1 | 2 | 3 | 4 | 5 |
|---|---|---|---|---|---|
| **09:00-09:30** | Registration | Module 3, session 6 | Module 3, session 10 | Module 4, session 14 | Module 5, session 19 |
| **09:30-10:30** | Module 1, session 1 | | | | |
| **10:30-11:00** | Coffee Break | | | | |
| **11:00-12:30** | Module 1, session 2 | Module 3, session 7 | Module 4, session 11 | Module 4, session 15 | Module 5, session 20 |
| **12:30-14:00** | Lunch break | | | | Adjourn |
| **14:00-15:30** | Module 2, sessions 3 and 4 | Module 3, session 8 | Module 4, session 12 | Module 4, session 16 | |
| **15:30-16:00** | Coffee break | | | | |
| **16:00-17:30** | Module 2, session 5 | Module 3, session 9 | Module 4, session 13 | Module 4, sessions 17, 18 | |

Typical examples of the contents of each session within the training modules are given below.

MODULE 1: Opening and introduction of the host State's national nuclear programme

Session 1: Opening
1. Welcome and opening remarks
2. Introduction of participants
3. Workshop objectives and agenda
4. IAEA project on "Capacity building on site safety review and assessment in embarking countries"

Session 2: Nuclear programmes and capacity in the host State
5. Overview of the host State's nuclear power programme
6. Status and progress of the national regulatory programme in the host State
7. Status and progress of the site selection and evaluation programme for nuclear installations

MODULE 2: General aspects of regulatory capacity for nuclear site safety review

Session 3: Pre-training self-assessment
8. Pre-training quiz on site safety review and assessment

Session 4: General aspects of nuclear site safety, part I
9. IAEA External Events Safety Section and its activities
10. Site safety review and licensing of nuclear installations

Session 5: General aspects of nuclear site safety, part II
11. Introduction to safety principles
12. Public involvement and acceptance

MODULE 3: Technical aspects of regulatory capacity for nuclear site safety review

Session 6: General technical aspects for nuclear site safety, part I
13. Natural hazards for site evaluation: overall aspects
14. Natural hazards for site evaluation: meteorological hazards

Session 7: General technical aspects for nuclear site safety, part II
15. Site survey and site selection for nuclear installations, part I
16. Site survey and site selection for nuclear installations, part II

Session 8: General technical aspects for nuclear site safety, part III

17. Site evaluation for nuclear installations, part I
18. Site evaluation for nuclear installations, part II

Session 9: Specific technical aspects for site safety review, part I

19. Site characteristics important to emergency planning and radiological environmental impact assessment (EIA), part I
20. Site characteristics important to emergency preparedness and radiological EIA II

Session 10: Specific technical aspects for site safety review, part II

21. Site characteristics important to emergency preparedness and radiological EIA, part III
22. Site characteristics important to emergency preparedness and radiological EIA, part IV

MODULE 4: Hands-on practice of site safety review of site characteristics important to emergency planning and radiological EIA

Session 11: Instructions and preparation for team practice

23. Instructions for team practice and team preparation
24. Introduction to sample documents

Session 12: SAR review

25. Review of SAR, part I

Session 13: Preparation of RAIs

26. Review of SAR, part II
27. Preparation of RAIs

Session 14: Answers to RAIs

28. Presentation of RAIs
29. Preparation of answers to RAIs

Session 15: RAIs Q&A meeting

30. Presentation of answers to RAIs and Q&As
31. Preparation of safety evaluation report

Session 16: Preparation of safety evaluation report

32. Preparation of safety evaluation report (continued)

Session 17: Presentation of safety evaluation report and public involvement

33. Presentation of safety evaluation report (team leaders) and Q&As for public acceptance

Session 18: Post-training self-assessment

34. Post-training quiz on site safety review and assessment

MODULE 5: Feedback and conclusion

Session 19: Results briefing and evaluation

35. Results briefing and evaluation of Module 4
36. Evaluation of the quiz results
37. Feedback and evaluation of the workshop

Session 20: Summary, discussion and closing

38. Summary of the workshop
39. Discussion on future activities
40. Concluding remarks

## I-4. QUESTIONNAIRES FOR TRAINING COURSE EVALUATION

An example of a set of questions for training course evaluation is as follows:

**Evaluation questionnaire: [*Title, Date, Venue, Event No.*]**

**Overall evaluation**

1. Your organization, position, role/tasks?

2. How satisfied are you with this overall training course?
   a. Extremely satisfied
   b. Very satisfied
   c. Somewhat satisfied
   d. Not so satisfied
   e. Not at all satisfied

3. How satisfied are you with the agenda and schedule of the training course?
   a. Extremely satisfied
   b. Very satisfied
   c. Somewhat satisfied
   d. Not so satisfied
   e. Not at all satisfied

4. How do you feel about the relationship between this training course and your work?
   a. Extremely relevant
   b. Very relevant
   c. Somewhat relevant
   d. Not so relevant
   e. Not at all relevant

5. What would be the optimal length of this kind of training course?
   a. 3 days
   b. 4 days
   c. 1 week
   d. Other (please specify)

6. Do you think that more time needs to be allocated to national presentations?
   a. Need to allocate more time
   b. Satisfied as it is
   c. Need to shorten the time

7. How likely is it that you would recommend this training course to others?
   a. Extremely likely
   b. Very likely
   c. Somewhat likely
   d. Not so likely
   e. Not at all likely

8. How well have you improved your knowledge and insight in the safety review and assessment of sites for nuclear installations?
   a. Extremely well
   b. Very well
   c. Somewhat well
   d. Not so well
   e. Not at all

9. How satisfied are you with the technical conditions (room, projection, sound, etc.) given during the training course?
   a. Extremely satisfied
   b. Very satisfied
   c. Somewhat satisfied
   d. Not so satisfied
   e. Not at all satisfied

10. Please write the numbers of the session(s) that you enjoyed most (up to 3 sessions).

11. Please write the numbers of the session(s) that you think most useful (up to 3 sessions).

12. Please provide any other suggestions regarding this training course.

**Evaluation of Module 1**

1. How satisfied are you with the Module 1 overall?
   a. Extremely satisfied
   b. Very satisfied
   c. Somewhat satisfied
   d. Not so satisfied
   e. Not at all satisfied

2. How satisfied are you with the contents of Module 1?
   a. Extremely satisfied
   b. Very satisfied
   c. Somewhat satisfied
   d. Not so satisfied
   e. Not at all satisfied

3. How relevant were the contents of Module 1 for your work?
   a. Extremely relevant
   b. Very relevant
   c. Somewhat relevant
   d. Not so relevant
   e. Not at all relevant

4. How do you feel about the amount of time that was allocated for the Module 1?
   a. Much too short
   b. Somewhat too short
   c. About the right amount of time
   d. Somewhat too long
   e. Much too long

5. Do you have any suggestions for the improvement of Module 1?

**(The same set of questions can be asked for the evaluation of Modules 2, 3, 4 and 5)**

## I-5. STRUCTURE OF A TRAINING COURSE SUMMARY REPORT

An example structure of a training course summary report is given below.

Cover page (title, date and venue of the training course, host organization, mission team members, report issue date)

1. Introduction
   a. Background
   b. Objectives of the training course
   c. Scope of the training course
   d. Overview of the activities

2. Self-assessment

3. Conduct of the training course
   a. Module 1
   b. Module 2
   c. Module 3
   d. Module 4
   e. Module 5

4. Conclusions (summary, achievements, feedback, recommendations, etc.)

5. IAEA reference documents

6. Appendices
   a. List of participants
   b. Agenda of training course
   c. Photo of the training course
   d. Presentation materials (link to the training course web page)
   e. Pre- and post-training quiz reports
   f. Report of training course evaluation questionnaire

## REFERENCES TO ANNEX I


[I-1] INTERNATIONAL ATOMIC ENERGY AGENCY, Basic Professional Training Course, https://gnssn.iaea.org/main/bptc, (2015).

[I-2] INTERNATIONAL ATOMIC ENERGY AGENCY, Regulatory Control of Nuclear Power Plants, Training Course Series No. 15, IAEA, Vienna (2002).

# ANNEX II EXAMPLES OF MEMBER STATE TRAINING PROGRAMMES ON THE REVIEW OF NUCLEAR INSTALLATION SITE SAFETY

## II-1.CAPACITY BUILDING PROGRAMME ON SITE SAFETY FOR NUCLEAR INSTALLATIONS IN THE KOREA INSTITUTE OF NUCLEAR SAFETY, REPUBLIC OF KOREA

### II-1.1. Human Capacity Building Programme for Embarking Countries

The Korea Institute of Nuclear Safety (KINS) developed its comprehensive Human Capacity Building Programme (HCB Programme) for embarking countries. This programme is also available in full or in part for training KINS staff. The HCB Programme is designed for trainees to develop and maintain the qualifications and competence needed for regulatory staff.

#### *II-1.1.1. Management of the Human Capacity Building Programme*

The management of the HCB Programme consists of the KINS Department of Education and Training, HCB Programme coordinator, the professors in charge of the programme, lecturers/mentors, and moderators. The roles and responsibilities assigned to them are as follows:

- KINS Department of Education and Training: Plans and implements the programme and acts as the main contact point.
- Professors: Develop the academic course of the programme and also act as academic advisers for the trainees. Experts are drawn from the fields of (i) general nuclear safety, (ii) safety analysis and accident management, (iii) nuclear power plant design and operation, (iv) electrical engineering, and (v) radiation protection.
- Lecturers/mentors: Develop the teaching materials. These are selected from the expert pool within KINS.
- Moderators: Manage the quality of the programme and assist the lecturers. Staff members of the KINS Department of Education and Training, specifically dedicated professors, lecturers or other qualified personnel can act as moderators of the lectures.

#### *II-1.1.2. Structure of the Human Capacity Building Programme*

The HCB Programme is divided into three levels with distinct educational goals.

Level 1 (Basic) is developed to educate the trainees for three months with the goal of teaching the fundamental principles and requirements of nuclear and radiation safety including IAEA safety standards and relevant Korean practices.

Level 2 (Intermediate) is developed to educate the trainees for five months with the goal of helping them to understand the specific regulatory technology and relevant requirements of various technical areas, such as siting, structural engineering, mechanical engineering, material engineering, nuclear systems, instrumentation and control, electrical engineering, safety analysis, severe accident and accident management, reactor and nuclear fuel, and radiation protection.

Level 3 (Advanced) is developed to train the trainees for nine months with the goal of practising licensing processes and regulatory activities through on-the-job training. On-the-job training is an extended form of practical training that provides an opportunity to learn how to perform complex and difficult tasks or to operate complex equipment under supervision.

The content and length of on-the-job training may vary significantly depending on the particular licensed activity and job requirements; trainees typically progress from shadowing job incumbents performing a particular task and observing how a task is performed, to performing the task under supervision, before carrying out the task unsupervised.

The Level 3 (Advanced) course is conducted through formal classroom lectures (including computer code exercises with personal laptop computers), case studies, group work, one-to-one mentoring, parallel work, discussions, and business inspection trips to practice review and assessment, inspection of nuclear safety and regulation.

The trainees' success is evaluated in two different aspects: technical achievement (70%) and attitude (30%). For the technical achievement, two examinations and a written assignment are given, while attitude is graded by attendance, participation and willingness to learn.

In the on-the-job training for inspection, trainees are divided into five different groups: (a) siting and structural engineering, (b) mechanical engineering, (c) radiation protection, (d) instrumentation and control and electrical engineering, and (e) reactor system engineering.

The on-the-job training for inspection is designed to have consistency by having the same mentors participating in both the classroom (mentoring) and in situ training. The classroom training is a prerequisite to the in situ training. In situ experience was highly valued in designing the programme. Trainees are able to directly observe the licensing of documents, the preparation of documentation for review and assessment, and in situ inspections. Finally, the course was designed to give insight into acceptance criteria during walkthroughs or visual inspections.

*II-1.1.3. Assessment, monitoring and evaluation of the Human Capacity Building Programme*

**Assessment of trainees**

Assessment is necessary to verify whether trainees have mastered the training objectives of the HCB Programme and acquired the competence needed to perform their jobs safely. Progress is assessed after each lesson or exercise, and again at the completion of the entire set of applicable lessons. Examinations to measure knowledge, and observation of performance during hands-on exercises to measure skill, are the common tools for these assessments. Testing is based on the training objectives and test items prepared in the design phase of training programme development.

The assessment results are communicated to the trainees as soon as possible after the assessment, along with any indication that they need additional training. Participants who fail a lesson examination during the HCB Programme, or whose performance during an exercise is poor, need to correct the problem area(s) before progressing in their training.

The different forms of training are assessed as follows:

(a) Knowledge-based learning: Written and oral examinations are normally used to evaluate the radiation safety knowledge workers have acquired.
(b) Skills-based learning: Assessment of practical exercises and on-the-job-training is the most direct way to determine whether workers have effectively learned the skills needed to perform their jobs safely.
(c) Practical exercises: The assessment of a practical exercise determines the worker's ability to perform a task or set of tasks by applying the appropriate knowledge and skills. The goal of the assessment is to confirm that the worker will be able to complete the tasks associated with the practical exercise without supervision. Practical exercise assessment is normally conducted during the exercise itself. For this reason, the instructor clearly identifies those observations, remarks and questions that are meant to assist the workers in mastering the training objectives of the exercise, and those that are intended as part of the graded assessment.
(d) On-the-job training: Assessment of on-the-job training is conducted after completion of the exercise. Before assessment of each activity, the worker receives all the information that would normally be available when performing that activity in real life. Then, during the assessment, the worker demonstrates competence in performing the activity using all the relevant procedures, tools and equipment.

**Monitoring and evaluation of the Human Capacity Building Programme**

The HCB Programme is monitored and evaluated in a number of ways. There are weekly meetings between the KINS Department of Education and Training and HCB Programme coordinator to discuss and plan the progress of the HCB Programme. Results from the meeting are documented and are used to improve the programme.

The KINS Department of Education and Training and the HCB Programme coordinator jointly developed a weekly questionnaire for trainees to evaluate the lectures, lecturers and programme in general. Each week, an online questionnaire link is provided to the trainees. The raw data results from the questionnaire is delivered to the HCB Programme coordinator every week and are used to adjust and modify the programme to meet the needs of the trainees.

To provide updated information on the HCB Programme, monthly reports are prepared containing information such as the progress of the training, evaluations and pictures.

The modules for different course levels provided in the HCB Programme are given in Table II-1.

TABLE II-1. COURSE LEVEL AND MODULE

| Level | Module |
|---|---|
| **1.** | *(※ General course provided to managers and staff of all technical areas)* |
| 0.1. | Introduction |
| 0.1.1. | Introduction of the HCB Programme |
| 0.1.2. | Governmental, legal and regulatory framework in the Republic of Korea |
| 0.1.3. | Licensing in the Republic of Korea |
| 1.1. | Overview of IAEA safety standards |
| 1.1.1. | Overview of IAEA safety standards (Safety Fundamentals, General Safety Requirements, Specific Safety Requirements) |
| 1.1.2. | Regulatory infrastructure of an NPP (IAEA Safety Standards Series No. SSG-16 (Rev. 1) [II-1]) |
| 1.1.3. | Governmental, Legal and Regulatory Framework for Safety (IAEA Safety Standards Series No. GSR Part 1 (Rev. 1) [II-2]) |
| 1.1.4. | Introduction of INSAG Report |
| 1.1.5. | Licensing Process |
| 1.1.6. | Overview of Nuclear Regulations |
| 1.1.7. | Construction of NPP |
| 1.1.8. | Commissioning of NPPs |
| 1.1.9. | Operation of NPPs |
| 1.1.10. | Safety Analysis Report |
| 1.1.11. | Review and Assessment of NPPs |
| 1.1.12. | Introduction to Nuclear Facilities |
| 1.2. | Fundamentals of NPP design |
| 1.2.1. | Safety Requirements of nuclear power plant design (IAEA Safety Standards Series No. SSR-2/1[II-3]) |
| 1.2.2. | Basic nuclear engineering and design |
| 1.2.3. | System design of nuclear steam supply system |
| 1.2.4. | System design of essential service feedwater system |
| 1.2.5. | System design of safety-related and auxiliary system |
| 1.2.6. | Steam and power conversion system |
| 1.2.7. | Plant protection, control and monitoring system |
| 1.2.8. | Siting and structural engineering |
| 1.2.9. | Mechanical and material engineering |
| 1.2.10. | Electrical power system |
| 1.2.11. | Technical specifications |
| 1.2.12. | Overview of APR 1400 |

TABLE II-1. COURSE LEVEL AND MODULE (cont.)

| Level | Module |
|---|---|
| 1.3. | Other topics |
| 1.3.1. | Safety requirements; commissioning and operation (IAEA Safety Standards Series No. SSR-2/2 (Rev. 1) [II-4]) |
| 1.3.2. | Licence amendment review |
| 1.3.3. | Regulatory inspection |
| 1.3.4. | Introduction to configuration management |
| 1.4. | Safety assessment of nuclear facilities |
| 1.4.1. | Safety assessment and verification (IAEA Safety Standards Series No. GSR Part 4 (Rev. 1)[II-5]) |
| 1.4.2. | Defence in depth and safety margins |
| 1.4.3. | Safety analysis; deterministic safety analysis, probabilistic safety assessment |
| 1.4.4. | Severe accident analysis |
| 1.4.5. | Accident management |
| 1.5. | Fundamentals of radiation safety |
| 1.5.1. | Radiation protection and radiation safety (IAEA Safety Standards Series No. GSR Part 3 [II-6]) |
| 1.5.2. | Radiation protection |
| 1.5.3. | Predisposal management of radioactive waste (IAEA Safety Standards Series No. GSR Part 5 [II-7]) |
| 1.5.4. | Radioactive waste management |
| 1.5.5. | Decommissioning of facilities (IAEA Safety Standards Series No. GSR Part 6 [II-8]) |
| 1.5.6. | Decommissioning |
| 1.5.7. | Radiological accidents and response |
| 1.6. | Emergency preparedness and response |
| 1.6.1. | Preparedness and response for a nuclear or radiological emergency (IAEA Safety Standards Series No. GSR Part 7 [II-9]) |
| 1.6.2. | Preparedness and response for a nuclear or radiological emergency |
| 1.6.3. | Emergency management system, roles and responsibilities |
| 1.7. | Other nuclear facilities |
| 1.7.1. | Overview of fuel cycle facilities |
| 1.7.2. | Overview of research reactors |
| 1.8. | Management system |
| 1.8.1. | IAEA Safety Standards Series No. GSR Part 2 [II-10] |
| 1.8.2. | Safety culture and its oversight; safety culture for regulatory body, Oversight of licensee's safety culture |
| 1.9. | Global nuclear safety regime |
| 1.10. | Quality assurance |
| **2.** | *(※ In-depth specific course provided to managers and staff of relevant technical areas)* |
| 2.1. | Site safety for nuclear installations |
| 2.1.1. | Site characteristics |
| 2.1.1.1. | Site characteristics (geography and demography)<br>I. Site location and description<br>II. Exclusion area<br>Population distribution |
| 2.1.1.2. | Site characteristics (human-induced hazard)<br>I. Investigation for hazard evaluation<br>II. Evaluation<br>Application |
| 2.1.1.3. | Site characteristics (meteorology)<br>I. Meteorological hazards<br>On-site meteorological monitoring programme |
| 2.1.1.4. | Site characteristics (hydrology)<br>I. Requirements in IAEA safety standards<br>II. Requirements in other countries<br>Hydrological evaluation |

TABLE II-1. COURSE LEVEL AND MODULE (cont.)

| Level | Module |
|---|---|
| 2.1.1.5. | Site characteristics (seismology)<br>I. Earthquake fundamentals<br>II. Earthquake monitoring in KINS<br>III. Seismic safety of NPP site in Republic of Korea |
| 2.1.1.6. | Site characteristics (geology and geotechnical engineering)<br>I. Overview<br>II. Investigation<br>III. Evaluation<br>IV. Foundation stability<br>V. Slope stability |
| 2.1.2. | Seismic safety<br>I. Regulatory requirements<br>II. Single-degree-of-freedom (SDOF) systems<br>III. Response spectra<br>IV. Response spectrum analysis of multi-degree-of-freedom (MDOF) systems<br>V. Seismic analysis methods<br>VI. Seismic classification<br>VII. Seismic design parameters<br>VIII. Seismic response analysis<br>IX. Seismic instrumentation and automatic seismic trip system<br>X. Seismic qualification methods<br>XI. Requirements for seismic qualification and documentation |
| 2.1.3. | Structural engineering |
| 2.2.~ 2.29. | Other than the site safety ( ※ *Exempted from the table)* |
| **3.** | *( ※ Hands-on practice and on-the-job training provided to managers and staff of all technical areas)* |
| 3.1. | Regulatory framework and licensing process |
| 3.1.1. | Introduction of Level 3 Overview |
| 3.1.2. | Regulatory framework and licensing process (including short exercise) |
| 3.1.3. | Review and assessment (including short exercise) |
| 3.1.4. | Regulatory inspection (pre-operational and periodic) |
| 3.1.5. | Resident inspection |
| 3.2. | Regulator training |
| 3.2.1. | Basic regulator training I (attitude, ethics, communication skills) |
| 3.2.2. | Basic regulator training II (drawing and layout (e.g. piping and instrumentation diagram (P&ID), logic diagrams), etc.) |
| 3.3. | Review and assessment of on-the-job training |
| 3.3.1. | Project management (risk and assessment) |
| 3.3.2. | Introduction of review and assessment programme |
| 3.3.3. | Outline of licensing review and assessment exercise (table-top) |
| 3.3.4. | Simulated on-the-job training for review and assessment process and activity (role play) |
| 3.3.5. | Review and assessment training and exercise mentoring) |
| 3.4. | Regulatory inspection on-the-job training |
| 3.4.1. | Simulated on-the-job training for quality assurance inspection (role play) |
| 3.4.2. | Project management (inspection) |
| 3.4.3. | Introduction of inspection training programme |
| 3.4.4. | Inspection (pre-operational and periodic inspection in Republic of Korea) |
| 3.4.5. | International inspection practice |
| 3.4.6. | Outline of inspection exercise (table-top) |
| 3.4.7. | Simulated on-the-job training for inspection (role play) |
| 3.4.8. | Inspection classroom training |
| 3.4.9. | Field inspection (mentoring) |

### II-1.2. Training Programme for KINS Staff

There are two types of regular training programmes for staff in KINS, one for newly recruited staff and the other for junior and senior staff. For the newcomer staff, a training programme of one to three months is provided and the training areas are as follows: nuclear safety fundamentals, NPP design, laws and regulations, licensing process and related documents, licensing review and assessment, regulatory inspection, quality assurance inspection, radiation protection, emergency preparedness and plan (EP&P), radioactive waste management, and APR-1400 simulator training.

The following are examples of lectures to be provided to new staff:

— Introduction of KINS;
— Nuclear safety fundamentals;
— Concept of radiation protection;
— Radiation protection;
— Safety concept and design of NPPs;
— Overview of NPP systems;
— Research reactors and nuclear fuel cycle facilities;
— Operational safety and operational guidelines of NPP;
— Investigation of events and accidents at NPP and root cause analysis;
— Quality assurance;
— Nuclear regulatory system;
— Laws and regulations;
— Probabilistic safety assessment and its application;
— Emergency preparedness and plan (EP&P);
— Regulation for radiation sources;
— Radioactive waste management;
— Licensing documents;
— Radiation exposure management;
— Safety review and assessment;
— Regulatory inspection;
— Nuclear accidents and events;
— Attitude of regulators;
— Integrated management system;
— Nuclear safety culture;
— Analysis of radioactive material;
— Industrial and medical radiation safety;
— Environmental radiation safety.

For junior and senior staff at KINS, it is mandatory to take intermediate and advanced courses for at least 40 hours per year. Each technical department also provides in-house training programmes annually and this training is evaluated as a department activity.

### II-1.3. Training programme for KINS Staff on site evaluation

A training programme for staff in each technical department of KINS, in general, consists of four categories such as fundamentals, intermediate, advanced and special lectures.

The training programme for site evaluation is as follows:

(1) Fundamentals
  1. Site Characteristics:
    • Introduction;
    • Geography and demography;
    • Human-induced hazards;

- Meteorology;
- Hydrology;
- Geology and seismology;
- Geotechnical engineering.

2. Seismic safety:
   - Introduction;
   - Regulatory requirements;
   - Seismic design;
   - Seismic qualification testing;
   - Seismic instrumentation and automatic seismic trip system.

(2) Intermediate

1. Site characteristics:
   - Geography and demography;
   - Human-induced hazards;
   - Hydrology;
   - Seismology;
   - Geology and geotechnical engineering.
2. Seismic Safety.

(3) Advanced

Advanced courses consist of two types: on-the-job training and on-the-job participation . On-the-job training is in general carried out as classroom type training and usually as a 1:1 mentoring approach with senior and new staff. On-the-job participation is performed during the regulatory inspection at the NPPs; similarly using the 1:1 mentoring approach with senior and new staff.

(4) Special lectures

Special lectures are provided by senior staff or experts from outside. Examples of the special lectures are as follows:

- Site-specific earthquake ground motions for NPPs;
- Recent slope stability analysis and design criteria;
- Design criteria improvement with advance estimation method of design wave under deep sea;
- State of the art technology for micro earthquake analysis;
- Characteristics and application of software for general earthquake analysis;
- Investigation of underground water at the second surface disposal facility and its modelling evaluation.

## II-2.CAPACITY BUILDING PROGRAMME ON THE REVIEW OF NUCLEAR INSTALLATION SITE SAFETY IN THE NUCLEAR REGULATORY AUTHORITY, TÜRKIYE

The objective of the training programme of the Turkish Nuclear Regulatory Authority (NDK) that the staff develop their administrative and technical qualifications and competencies within the scope of a training management system (see Refs [II-11] to [III-15]. Continuing training plans are made, taking into account the needs of the NDK and in accordance with the personnel policy, and are approved by the NDK Board.

An annual training programme is prepared in accordance with the training plan. Training can take place in the NDK or in other organizations in Türkiye or abroad, depending on the needs and type of training. The annual training programme is prepared by the NDK Department of Support Services with input from other units, and is approved by the NDK President.

Training is given to personnel in order to develop skills related to their duties, increase their practical and theoretical knowledge, train them in service and ensure efficiency, effectiveness and high quality.

### II-2.1. Training catalogue

Tables II-2 and II-3 show lists of trainings on nuclear technology and nuclear regulation, respectively, offered to NDK staff.

TABLE II-2. TRAINING PROGRAMME ON NUCLEAR TECHNOLOGY FOR STAFF OF THE NDK

| NUCLEAR TECHNOLOGY TRAINING | |
|---|---|
| OBJECTIVE | To inform the relevant personnel about the usage areas of nuclear technology, such as energy, medicine and industry, and the developments in Türkiye and in the world. |
| CONTENT | 1. Development process of nuclear technology<br>2. Uses of nuclear technology (energy, medicine, agriculture and industry)<br>3. World energy demand and nuclear power<br>4. Nuclear energy in Turkey and the world<br>5. International Atomic Energy Agency (IAEA)<br>6. Climate change and the impact of energy resources<br>7. Nuclear power plant and nuclear accidents<br>8. Nuclear technology and radiation safety<br>9. Other uses of nuclear technology<br>10. The future of nuclear technology<br>11. International nuclear treaties and conventions |
| TARGET AUDIENCE | Relevant staff of the NDK |
| DURATION | According to needs. |
| FORMAT | Online or in person |
| TRAINING METHODS | Presentations, case studies, Q&A, group work, brainstorming, computer aided teaching |
| TRAINING MATERIAL AND EQUIPMENT | Computer (with necessary programs and systems), projector, relevant legislation, lecture notes, training forms. |
| ASSESSMENT | At the end of the training, participants will be given a certificate of participation.<br>Presenter satisfaction questionnaire<br>Training evaluation questionnaire |

TABLE II-3. TRAINING PROGRAMME ON NUCLEAR REGULATION FOR STAFF OF THE NDK

| NUCLEAR REGULATION TRAINING | |
|---|---|
| OBJECTIVE | To provide staff with regulatory qualifications, to develop these qualifications further and to inform trainees about international regulatory structures and approaches. |
| CONTENT | 1. Characteristics of regulatory bodies, scopes of duty<br>2. Requirements to become a nuclear regulatory authority<br>3. Regulatory activities<br>4. Safety and security culture<br>5. Regulatory stances, attitudes and behaviours to authorized persons<br>6. Regulatory approaches (performance-based, prescriptive)<br>7. International Atomic Energy Agency (IAEA) and the Nuclear Energy Agency of the Organisation for Economic Co-operation and Development (OECD/NEA)<br>8. United Nations Scientific Committee on the Effects of Atomic Radiation (UNSCEAR) and International Commission on Radiological Protection (ICRP)<br>9. Other regulatory associations, such as the European Nuclear Safety Regulators Group (ENSREG) and Western European Nuclear Regulators Association (WENRA)<br>10. International conventions and agreements<br>11. Bilateral cooperation agreements |
| REGULATORY APPROACH TRAINING | |
| TARGET AUDIENCE | Relevant staff of the NDK |
| DURATION | According to needs |
| FORMAT | Online or in person |
| TRAINING METHODS | Presentation, case study, Q&A, group work, brainstorming, computer aided teaching |
| TRAINING MATERIAL AND EQUIPMENT | Computer (with necessary programs and systems), projector, relevant legislation, lecture notes, training forms |
| ASSESSMENT | At the end of the training, participants will be given a Certificate of Participation.<br>Presenter satisfaction questionnaire<br>Training evaluation questionnaire |

### II-2.2. Training types

*II-2.2.1. Training programmes specifically for newly hired technical staff*

New technical staff at the NDK complete a preparatory training programme consisting of three stages.

**Stage 1: Initial training**

The aim of this training is to ensure that newly recruited technical staff have information about the institutional functioning, duties and responsibilities of the NDK, including:

— General information about administrative and financial management within the scope of the NDK, and about current administrative legislation and circulars;
— General information about the technical duties and responsibilities of the NDK;
— The importance given to safety and information security in NDK activities.

### Stage 2: Basic nuclear energy training

The aim of this training is to provide basic nuclear energy training to staff working at the NDK who have not received nuclear energy training in their undergraduate or graduate education. The aim of the training is to equip staff with basic knowledge about nuclear reactors, reactor engineering, radiation protection, reactor physics, reactor design, accident analysis and emergency management.

A typical example of the structure and contents of the training programme is given below.

1) Nuclear reactor basic principles:
   - Radioactivity and nuclear physics overview;
   - Introduction to nuclear physics;
   - Introduction to reactor physics;
   - Basics of thermal hydraulics.
2) Radiation protection:
   - Units and special parameters;
   - Natural radiation;
   - Detectors;
   - Stochastic and deterministic effects of radiation;
   - Principles of radiation protection;
   - Radiation protection applications in nuclear power plants, measures taken.
3) Nuclear power plant sites:
   - Nuclear power plant site selection;
   - Nuclear power plant site evaluation.
4) Nuclear reactor design:
   - Reactor types;
   - VVER-1000 and VVER-1200 technology;
   - Concept of defence in depth;
   - General facility design;
   - Facility system design requirements.
5) Classification of structures, systems and components:
   - Safety classification;
   - Quality classification;
   - Seismic classification.
6) Quality management in nuclear facilities:
   - Quality management.
7) Deterministic accident analysis:
   - Initiating events and their classification;
   - Acceptance criteria;
   - Selection of initial and boundary conditions;
   - Loss of coolant accidents.
8) Probabilistic safety assessment:
   - Basic concepts;
   - General methodology;
   - Level 1 probabilistic safety assessment;
   - Level 2 probabilistic safety assessment;
   - Level 3 probabilistic safety assessment.
9) Operational conditions:
   - Basic concepts;
   - Safety system settings;
   - Limiting conditions for normal operation.
10) Three Mile Island, Chernobyl and Fukushima accidents and lessons learned

An example schedule for a one-week basic nuclear energy training course is presented in Table II-4.

TABLE II-4. EXAMPLE SCHEDULE FOR A ONE-WEEK BASIC NUCLEAR ENERGY TRAINING COURSE

| | **Monday** | **Tuesday** | **Wednesday** | **Thursday** | **Friday** |
|---|---|---|---|---|---|
| | Nuclear reactor basic principles | NPP sites | (cont.) | Deterministic accident analysis | Three Mile Island, Chernobyl and Fukushima accidents and lessons learned |
| | (cont.) | (cont.) | (cont.) | (cont.) | (cont.) |
| | (cont.) | Nuclear reactor design | Classification of structures, systems and components | Probabilistic safety assessment | |
| | (cont.) | (cont.) | (cont.) | (cont.) | |
| | Radiation protection | (cont.) | Quality management in nuclear facilities | Operational conditions | |
| | (cont.) | (cont.) | (cont.) | (cont.) | |

**Stage 3: Basic nuclear safety and security training**

This training is aimed at providing basic information on nuclear safety and security issues within the framework of regulatory activities.

A typical example of the structure and contents of the training programme is given below.

1) Regulatory authority functions and duties:
   — Legislation preparation;
   — Review and evaluation;
   — Authorization;
   Audit and sanctions;
   — Regulatory personnel definition and ethical values.
2) Radiation protection.
3) Safety:
   — Safety culture;
   — Nuclear safety;
   — Radiation safety;
   — Transport safety;
   — Radioactive waste safety.
4) Security:
   — Security culture and international framework (non-proliferation, safeguards, physical protection, import/export control).
5) Legislation:
   — NDK legislation;
   — IAEA documents;
   — Other (ICRP, European Utility Requirements (EUR), WENRA, international organizations and regulations).
6) Nuclear safety of facilities:
   — General principles of safety;
   — Basic approaches to safety (Integrated Management System (IMS), graded approach, classifications, principles such as redundancy);

— Nuclear power plant safety principles;
— Research reactor safety principles;
— Radioactive waste safety principles.

7) Nuclear security of facilities:
— Physical protection;
— Import/export control;
— Safeguards;
— Non-proliferation.

8) Emergency planning.
9) XXX project management plan and licensing process.
10) YYY project management plan and licensing process.

An example schedule for a one-week basic nuclear safety and security training course is presented in Table II-5.

TABLE II-5. EXAMPLE SCHEDULE FOR A ONE-WEEK BASIC NUCLEAR SAFETY AND SECURITY TRAINING COURSE

| | **Monday** | **Tuesday** | **Wednesday** | **Thursday** | **Friday** |
|---|---|---|---|---|---|
| | Regulatory authority functions and duties | Security | Nuclear safety of facilities | Nuclear safety of facilities | Nuclear safety of facilities |
| | Safety culture and nuclear safety | Legislation | (cont.) | (cont.) | Emergency planning |
| | Radiation safety and radiation protection | (cont.) | (cont.) | (cont.) | XXX project management plan and licensing process |
| | Transport safety | (cont.) | (cont.) | (cont.) | YYY project management plan and licensing process |
| | Radiation protection | Nuclear reactor design | Quality management in nuclear facilities | Operational conditions | |
| | (cont.) | (cont.) | (cont.) | (cont.) | |

*II-2.2.2. Training specifically for trainee inspectors*

This training is aimed at providing basic information to trainee nuclear safety inspectors. The following areas are covered in the training:

— Nuclear regulatory authority law;
— Decree on licensing of nuclear installations;
— Regulation on nuclear safety inspections and sanctions;
— Principles of professional ethical behaviour to be followed by inspectors;
— Directive on the authorization of nuclear safety inspectors;
— Directive on nuclear safety inspections;
— Inspection requirements;
— Inspection ethics and culture;
— Teamwork;

— Problem solving;
— Inspection management and methods;
— Management systems;
— Radiation protection;
— Findings management.

*II-2.2.3. Training specifically for the site safety assessment team*

The aims of this training are (a) to share information and experiences on reviewing and assessing NPPs on topics related to site selection and evaluation, (b) to disseminate the lessons learned from external hazards assessment and to discuss their impact on the revisions to the IAEA safety standards, and (c) to assist with the identification of actions to be undertaken. An example of a weekly schedule for a nuclear site safety training course is given in Table II-6.

TABLE II-6. AN EXAMPLE OF TRAINING COURSE FOR NUCLEAR SITE SAFETY

| Session | Day 1 | Day 2 | Day 3 | Day 4 |
|---|---|---|---|---|
| 1 | Licensing and site evaluation process, related regulations and standards (Lecturer xx) | Meteorological investigations in site evaluation for NPPs (Lecturer xx) | Effects of ecology, oceanography studies and environmental monitoring in site evaluation for NPPs (Lecturer xx) | Hydrological hazards in site evaluation for nuclear installations (floods), tsunamis – tsunami hazard assessment (Lecturer xx) |
| 2 | Siting (site survey and site selection) for NPPs (Lecturer xx) | Population considerations, radiological impact and demonstration of the feasibility of emergency plan (Lecturer xx) | Dispersion of radionuclides in groundwater systems: hydrogeological site evaluation for NPPs (Lecturer xx) | Environmental impact assessment for NPPs (Lecturer xx) |
| 3 | Tectonic framework and seismicity (Lecturer xx) | Ranking criteria and pre-evaluation in siting (Lecturer xx) | Geotechnical aspects and foundation safety of NPPs (Lecturer xx) | Siting activities for NPPs (Lecturer xx) |
| 4 | Seismic hazard assessment for NPP sites (Lecturer xx) | Site inspections of NPPs (Lecturer xx) | Human induced events in site evaluation (Lecturer xx) | General information on existing NPP sites (Lecturer xx) |

## II-3.CAPACITY BUILDING PROGRAMME ON THE REVIEW OF NUCLEAR INSTALLATION SITE SAFETY IN THE FEDERAL AUTHORITY FOR NUCLEAR REGULATION, UNITED ARAB EMIRATES

One of the top priorities of the Federal Authority for Nuclear Regulation (FANR) is to develop the skills and competencies of its staff to ensure a sustainable regulatory framework that is worldclass. Programmes have been developed with the strategic aim to provide FANR staff with the knowledge and skills to regulate the nuclear programme in the United Arab Emirates (UAE).

FANR's capacity building strategy comprises two main tracks: 1) staffing by senior expatriate staff to deal with the short-to-medium term regulatory needs; 2) competence development of UAE nationals to ensure long-term sustainability. FANR holds a strong cadre of senior and experienced professionals in the fields of nuclear safety and radiation protection. These experienced professionals assisted in developing the regulatory infrastructure including the regulatory framework as seen today. The competence development of young Emiratis is fulfilled through structured formal education programmes, such as scholarships provided to high performing undergraduate and postgraduate students in the field of nuclear engineering and radiation safety. Vocational on-the-job programmes such as the Inspector Qualification Programme and the FANR Emergency Preparedness and Response Organization Programme are fundamental training programmes needed for all employees engaged in the core regulatory activities at FANR.

In line with its vision and mission, FANR has maintained its focus on developing relevant competence for newly recruited staff; whilst maintaining areas of expertise for the existing workforce. During the latter part of 2015, the education and training team, in consultation with the IAEA and various best practice regulators around the world, embarked on a competency building programme to establish a robust framework within the Operations Unit. This competency development framework is the foundation for continuously and consistently ensuring a competent cadre, and quality assurance measurement for sustainability and growth and is now reflected in a training plan for each job position in FANR. The FANR training plans are also based on the competence model described in IAEA Safety Reports Series No. 79 [II-16] which is distributed on a quadrant structure: Quadrant 1 contains the competences related to the legal, regulatory and organizational basis; Quadrant 2, the competences related to technical disciplines; Quadrant 3, the competences related to a regulatory body's practices; and Quadrant 4, the personal and behavioural competences.

Each quadrant comprises a set of quadrant competence areas (QA), and each of these quadrant competence areas comprises a set of specific competences referred to as KSAs as shown in Fig. 2.

FANR established a Leadership and Management Development programme for employees to become better leaders and contributors within FANR, as well as valuable contributors to FANR's broader regulatory and transparency goals. FANR's leadership competency framework, designed in 2019, serves as the baseline for all leadership programmes. FANR leaders participated in several leadership activities such as the Continuous Leadership Development programme, various UAE leadership programmes, the Management Senior Reactor Operator Programme and IAEA International School on Nuclear and Radiological Leadership for Safety. In addition to these programmes, FANR also engaged its managers at all levels in a series of customized training modules, including 'Emotional Intelligence for Leaders', 'Team Leadership', 'Strategic Thinking', 'People Performance and Management', 'Empowerment and Delegation' and 'She is a Leader'.

During the siting and construction phase of the UAE NPP, FANR initiated its 'Nuclear Reactor Assessor Training Programme' to provide highly specialized training on topics such as siting. Technical support organizations were used to lead some of the training activities.

### II-3.1. Training format

The training courses are held in a variety of formats, which are described in Table II-7.

TABLE II-7. DESCRIPTION OF THE VARIOUS TRAINING FORMATS

| TRAINING FORMAT | DESCRIPTION |
|---|---|
| CLASSROOM TEACHING | Information is conveyed to a group that is led by an instructor. This training allows a large quantity of information to be presented and occurs when a large group of learners are scheduled for training at the same time. This format may involve lectures, demonstrations and discussions. |
| WORKSHOP | Course material is delivered in a hands-on environment. This training is useful when tasks and skills need hands-on practice. The work conditions are simulated to give the trainees experience in a realistic training setting. |

TABLE II-7. DESCRIPTION OF THE VARIOUS TRAINING FORMATS (cont.)

| TRAINING FORMAT | DESCRIPTION |
|---|---|
| ON THE JOB TRAINING | Training is conducted by qualified personnel in the actual work environment. This allows the trainee to make a positive contribution to the mission of the company while being trained. In this training setting, trainees achieve learning objectives through training conducted in the job environment. |
| FLIPPED CLASSROOM | Instructional strategy and a type of blended learning that reverses the traditional learning environment by delivering instructional content, often online, outside of the classroom. It moves activities, including those that may have traditionally been considered homework, into the classroom. In a flipped classroom, trainees watch online lectures, collaborate in online discussions, or carry out research at home prior to classroom attendance, thus assimilating the concepts before, to later on engage in discussions in the classroom with the guidance of a mentor or experienced person. |
| SELF-STUDY | A training setting without a full-time instructor in which the course content is provided through the training materials. |

On-the job training is usually guided or supervised by a mentor, who may evaluate its accomplishment by the following means:

- Performance: The trainee performs the task(s) under the supervision of the mentor, gaining experience by doing.
- Observation: The trainee acquires the necessary knowledge and skills by observing the execution of the relevant tasks.
- Discussion: Certain tasks might not offer the possibility to be performed or observed in normal circumstances. In these cases, knowledge and skills can be transferred through discussions with more senior workers and/or the mentor.

### II-3.2. Training courses available

The following tables show details of the courses available at FANR, including the overall purpose of the course, the learning objectives, the training format and the estimated duration. The common training courses and specific training courses are shown in Tables II-8 and II-9, respectively.

TABLE II-8. COMMON TRAINING COURSES

| COURSE | LEARNING OBJECTIVES | OVERALL PURPOSE OF THE COURSE | TRAINING FORMAT AND DURATION |
|---|---|---|---|
| UAE GENERAL LEGISLATION | Outline the UAE legislative structure and relevant aspects of the following legal fields:<br>- Administration legislation<br>- Individual rights<br>- Criminal law<br>- Public health and safety<br>- Labour health and safety<br>- Environmental protection | Provide knowledge and understanding of the national legislative framework in the different non-nuclear fields | Flipped classroom (16 hours) |
| | Interpret and apply the provisions of the above laws to a set of limited situations, under guidance | | |

TABLE II-8. COMMON TRAINING COURSES (cont.)

| COURSE | LEARNING OBJECTIVES | OVERALL PURPOSE OF THE COURSE | TRAINING FORMAT AND DURATION |
|---|---|---|---|
| UAE NUCLEAR LEGISLATION | Outline the legislative structure and relevant aspects of UAE laws related to radiation and nuclear safety | Acquire basic knowledge of UAE laws and regulations related to radiation and nuclear safety | Flipped classroom (16 hours) |
| | Describe the main relevant legal aspects of nuclear legislation:<br>- UAE nuclear law and liability<br>Special emphasis on the significance of Article 4 (Nuclear Law 6/2009) as the driver of all FANR regulations and guides<br>Licenses conceded by the Regulator | | |
| | Describe the rights of all stakeholders affected directly or indirectly by the provisions of the legal basis of FANR | | |
| | Interpret and apply the provisions of the above laws to a set of limited situations, under guidance | | |
| INTERNATIONAL NUCLEAR INSTRUMENTS | Define and explain relevant international instruments and documentation such as:<br>- Convention on Nuclear Safety;<br>- Joint Convention on the Safety of Spent Fuel Management and on the Safety of Radioactive Waste Management;<br>- Convention on Early Notification of a Nuclear Accident;<br>- Convention on Assistance in the Case of a Nuclear Accident or Radiological Emergency;<br>- Code of Conduct on the Safety and Security of Radioactive Sources;<br>- IAEA safety standards (regarding the safety fundamentals and general structure and role of these standards). | Obtain expertise in the regulatory framework regarding nuclear conventions and instruments at international level | Flipped classroom (16 hours) |
| | Interpret and apply the provisions of the above laws to a set of limited situations, under guidance | | |
| FUNDAMENTALS OF NUCLEAR SAFETY AND SAFETY STANDARDS | Define safety and its importance within nuclear activities and facilities | Introduce the concept of safety, how it is present in FANR's activities, how standards define safety criteria and an overview of how safety can be managed and ways to evaluate it. | Classroom (16 hours) |
| | Enunciate the fundamental safety objective | | |
| | Describe the IAEA's fundamental safety objective | | |
| | List the IAEA ten fundamental safety principles [II-17] and explain their importance to FANR's overall mission | | |
| | Describe the national safety standards | | |
| | Describe the IAEA's safety standards | | |
| | Identify applicable industrial standards (e.g.: the American Society of Mechanical Engineers (ASME), Institute of Electrical and Electronics Engineers (IEEE)) and why they are relevant to FANR's activity | | |
| | Outline the safety culture within the nuclear regulatory bodies and within licensees | | |

TABLE II-8. COMMON TRAINING COURSES (cont.)

| COURSE | LEARNING OBJECTIVES | OVERALL PURPOSE OF THE COURSE | TRAINING FORMAT AND DURATION |
|---|---|---|---|
| FUNDAMENTALS OF NUCLEAR SAFETY AND SAFETY STANDARDS (cont.) | List the main aspects of safety management: Hazard assessment, radiological risk evaluation, other risks, engineering safeguards, emergency preparedness | (cont.) | (cont.) |
| | State the fundamentals of safe operation evaluation and the need for review and inspection | | |
| | Detail the main aspects of safety management: Hazard assessment, radiological risk evaluation, other risks, engineering safeguards, emergency preparedness | | |
| | Explain the basis for the technical requirements in the area of nuclear safety for siting, design, construction, commissioning, operation, decommissioning and waste management of nuclear facilities/activities and devices | | |
| FUNDAMENTALS OF NUCLEAR SECURITY | List and recognize the objective and essential elements of a State's nuclear security regime | Introduce the concept of nuclear security, its principles, and main functions within the industry | E-learning (4 hours) |
| | State the IAEA's 12 essential elements of a State's nuclear security regime [II-18] | | |
| | Explain the UAE nuclear security regime | | |
| | Become familiar the content of the IAEA Nuclear Security Series and determine their usability in the UAE nuclear security regime | | |
| | Overview the main aspects of the planning, preparedness for and response to nuclear security events | | |
| | Identify the relationship between FANR and the UAE governmental organizations for physical protection, and understand their responsibilities | | |
| | List the design, operation and security controls of approved nuclear devices and equipment | | |
| | Outline FANR policies and principles for international cooperation | | |
| FANR MANAGEMENT SYSTEM FUNDAMENTALS | List the objectives to be covered by FANR's management system. | Present FANR's management system: what it is used for, how it is operated and how to assess and improve its effectiveness to support FANR's main activities. | E-learning (4 hours) |
| | Describe the structure, scope and operation of the system. | | |
| | Identify the procedures to measure and assess its effectiveness. | | |
| | Relate areas in which it may be improved with time and use according to FANR's approach. | | |
| FANR INTEGRATED MANAGEMENT SYSTEM | Define FANR's IMS | Describe the IMS and how it can be used to coordinate and control FANR's activities (including working with external organizations) | Classroom (6 hours) |
| | Describe the processes related to the IMS. | | |
| | Become familiar with IAEA Safety Standards Series Nos GSR Part 2 [II-19] and GS-G-3.1 [II-20]) which provide recommendations on management systems | | |

TABLE II-8. COMMON TRAINING COURSES (cont.)

| COURSE | LEARNING OBJECTIVES | OVERALL PURPOSE OF THE COURSE | TRAINING FORMAT AND DURATION |
|---|---|---|---|
| FANR INTEGRATED MANAGEMENT SYSTEM (cont.) | Explain the evolution of the IMS from its beginnings in quality control, quality assurance and total quality management through to its present form and the role of the IAEA in its continuous improvement | Describe the IMS and how it can be used to coordinate and control FANR's activities (including working with external organizations) | Classroom (6 hours) |
| | Describe how FANR's IMS is designed to consider the needs of all its stakeholders as well as organizational clients and external customers | | |
| | Describe FANR's approaches to measure, assess and improve the effectiveness of the IMS | | |
| | Describe the administrative and legal procedures associated to the internal management of documents | | |
| | Identify what indicators make up a useful filing system | | |
| | Identify how to outline methodologies to make record keeping systematic and effective | | |
| | Describe established data systems and procedures for storing, arranging, indexing and classifying records | | |
| | Practise a set of common operations | | |
| | Explain the graded approach to the implementation of the management system. | | |
| FUNDAMENTALS OF REVIEW AND ASSESSMENT | List the objectives of FANR's review and assessment processes | Introduce the basis of why and what FANR reviews and assesses across its departments (a review of procedures on how to do it is included although these and their associated standards to review against are detailed in later specific departmental courses) | Classroom (8 hours) |
| | Describe the contents of the main documents to be reviewed by FANR, (e.g. SAR, preliminary safety analysis report, quality assurance) specifically those of a safety analysis report | | |
| | List the main areas of review at each corresponding lifetime stage (siting, construction, design, operation, and decommissioning) | | |
| | Outline the network of FANR's Review Instructions (FANR-RI-xxx) | | |
| | Describe how technical support organization can assist FANR in review and assurance tasks, give examples of some applicable areas. | | |
| | Describe how review and assessment conclude in evaluation reports to judge the adequacy of the reviewed areas and how such reports are comprehensively structured | | |
| | Describe how review areas may be concluded, for example through requests [II-16] for additional information (RAIs) or conditional acceptance | | |
| | Explain the principles of targeting, proportionality and consistency in review and assessment tasks | | |
| | Define the concepts of uncertainties and conservatism in review and assessment | | |
| | Discuss how the review and assessment process informs and complements other processes such as inspection | | |
| | List specific lessons learned from the accident at the Fukushima Daiichi NPP and how they have been implemented in the industry | | |

TABLE II-8. COMMON TRAINING COURSES (cont.)

| COURSE | LEARNING OBJECTIVES | OVERALL PURPOSE OF THE COURSE | TRAINING FORMAT AND DURATION |
|---|---|---|---|
| ROOT CAUSE ANALYSIS | Explain the objectives and areas of application for root cause analysis (RCA) | Describe and apply RCA methods for the investigation of significant events at nuclear facilities | Flipped classroom (16 hours) |
| | Describe and analyse the main root cause analysis methods:<br>- Event and conditional factors analysis<br>- Fault tree analysis<br>- Management Oversight and Risk Tree (MORT) analysis<br>- Pareto analysis<br>- Barrier analysis | | |
| | Apply MORT analysis and other RCA techniques to conduct root cause analysis of several case study scenarios | | |
| | Describe quality assurance (QA) procedures applicable to internal audit | | |
| NUCLEAR SCIENCE AND TECHNOLOGY FUNDAMENTALS | Establish the basics behind nuclear physics and radiation, nuclear power plant technologies, nuclear fuel, nuclear security and safety, radiation protection, radioactive waste management, other applications of nuclear technology, environmental and economic aspects of nuclear technology. | Provide all workers with the fundamentals of nuclear energy and its uses | Classroom (8 hours) |
| FUNDAMENTALS OF RADIATION PROTECTION | Explain the concepts of:<br>- Ionizing radiation and its hazards<br>- Principles of health physics | Provide knowledge of the fundamental theoretical concepts of nuclear energy, what radiation is, how it is measured, how it interacts with matter, how to protect from it, and its applications in the industry | E-learning (6 hours) |
| | Outline the basic concepts of radiation protection for planned exposure situations, occupational exposure, medical exposure, public exposure, existing exposure situations and emergency exposure situations in relation to nuclear material and radioactive Sources | | |
| | List teaching styles that can be included in a learning programme which contains the characteristics and main values from the current students | | |
| | Explain direct relation between communication and training | | |
| | Identify communication factors which affect formative action | | |
| | Enumerate communication axioms according to P. Watzlawick et al. [II-21] | | |
| | Describe the importance of distortion in the communication process | | |
| | Enumerate and explain interpersonal communication keys | | |
| | Describe how the audience has to be treated in formative environments | | |
| | List and explain different techniques directed to improve communication abilities of the teachers | | |
| | Define type of questions and explain when they have to be used during a formative session | | |
| | Describe the main didactic methods that can be use in a formative session | | |
| | Use teaching methods which reflect the students' learning principles | | |

TABLE II-8. COMMON TRAINING COURSES (cont.)

| COURSE | LEARNING OBJECTIVES | OVERALL PURPOSE OF THE COURSE | TRAINING FORMAT AND DURATION |
|---|---|---|---|
| FUNDAMENTALS OF RADIATION PROTECTION (cont.) | Describe students' behaviours which can affect formative sessions and how to manage difficult situations in the training setting | (cont.) | (cont.) |
| | Identify the cause of most common fears in the instructor and describe techniques to overcome it | | |
| COACHING | Apply the skills related to emotional intelligence while using coaching and correction tools | Acquire the precise skills to identify, carry out and generate in the teams' expected behaviours through the coaching and correction techniques | Classroom (8 hours) |
| | Identify the appropriate situations to use positive reinforcement, feedback, coaching and correction techniques | | |
| | List coaching skills necessary to perform a good coaching process | | |
| | Describe the steps to perform a coaching or correction process | | |
| | Practice coaching and correcting techniques through role playing | | |

TABLE II-9. SPECIFIC TRAINING COURSES

| COURSE | LEARNING OBJECTIVES | OVERALL PURPOSE OF THE COURSE | TRAINING FORMAT AND DURATION |
|---|---|---|---|
| TRAINING ON LICENCE BASIS, DESIGN, OPERATION DOCUMENTATION OF NPPS AND TECHNICAL STANDARDS | Explain how the design and operation of NPP's are related to safety | Identify and analyse all the detailed areas of technical requirements associated with an NPP operating licence and how they are related with meeting safety objectives | Flipped classroom (32 hours) |
| | Describe the classification of structures, systems and components according to safety class and their requirements | | |
| | Associate design parameters with the applicable technical standards of their specific area | | |
| | Identify the main documents associated with the design and explain their function and content (e.g. process and instrumentation diagrams, control systems logic, protection systems, safety systems functionality, equipment specifications) | | |
| | Describe the main documents associated with NPP operation: operating procedures (e.g. general, system specific, abnormal, failure, emergency), operational limits and the methodology upon which they have been developed | | |
| | Explain the process by which design and operation basis, requirements and limits lead to the establishment of licence conditions | | |
| | Match the design and operation technical requirements to the licence conditions | | |
| UNCERTAINTY ANALYSIS | Define uncertainty and identify calculations within the safety case in which it inherently appears | | |
| | Describe the fundamentals statistical methods to evaluate and manage uncertainty | | |
| | Explain how to delimit its effects and how it may impact the compliance of overall safety objectives | | |
| | Explain how to consider uncertainties in the process of developing license conditions (e.g. operational limits) | | |

TABLE II-9. SPECIFIC TRAINING COURSES (cont.)

| COURSE | LEARNING OBJECTIVES | OVERALL PURPOSE OF THE COURSE | TRAINING FORMAT AND DURATION |
|---|---|---|---|
| FUNDAMENTALS OF SAFETY ANALYSIS METHODS | Define the concept of risk and state the principles of safety management | Explain the main safety analysis methods which will be used to support the safety cases in licence applications or modifications, and which will have to be reviewed and assessed by FANR | Classroom (16 hours) |
| | Describe different types of risk assessments: quantitative, qualitative, generic | | |
| | Understand what makes a risk assessment suitable and sufficient | | |
| | Describe the different philosophies to analyse safety of facilities and activities, including the basis and methods of deterministic and probabilistic approaches, including:<br>- Defining design basis accidents and beyond design basis accidents<br>- Describe how postulated initiating events are derived and how probabilities (frequencies) are assigned. | | |
| | Explain the concept of defence in depth | | |
| | Analyse systematic hazard and operability studies | | |
| | Give examples of risk controls put in place and how to mitigate them; describe how to evaluate their performance and adequacy | | |
| REVIEW AND ASSESSMENT REQUIREMENTS AND METHODS: NUCLEAR SAFETY SPECIFIC | Identify the different areas that FANR will review and assess to evaluate the safety of NPPs for each stage in their lifetime, from siting to decommissioning | Extend the review and assessment fundamentals to detailed requirements and related to nuclear safety of NPPs and radiation facilities. Identify the areas of review, contents of safety supporting documents to be reviewed and associated assessment methodologies, specific standards including, but not limited to, ASME, IEEE, IAEA, NRC and FANR review procedures. | Classroom (18 hours) |
| | Identify the documents and specific sections that conform to the licensee's safety case, which will be reviewed by FANR (technical documents, management system, QA, specific analysis reports) | | |
| | Match each review area to the associated safety requirements and applicable FANR regulations and guides | | |
| | Associate the specific groups of standards that apply to each review area (e.g. ASME, IEEE, IAEA, NRC) | | |
| | Describe the assessment methodology applicable for the safety analysis of each field/area (including those assigned to specialized TSOs) in order to evaluate compliance with regulations or fulfilment of safety objectives | | |
| | List cases where FANR review and assessment is also necessary, other than licence application processes: for example design modifications, operation, and maintenance procedure modifications | | |
| | Distinguish the applicable review areas that also apply to the nuclear safety of other radiation facilities (e.g. radiation waste processing plants) | | |

TABLE II-9. SPECIFIC TRAINING COURSES (cont.)

| COURSE | LEARNING OBJECTIVES | OVERALL PURPOSE OF THE COURSE | TRAINING FORMAT AND DURATION |
|---|---|---|---|
| REVIEW AND ASSESSMENT REQUIREMENTS AND METHODS: NUCLEAR SAFETY SPECIFIC (cont.) | Distinguish and explain FANR's review and assessment procedures (Review Instructions: FANR-RI-xxx) which guide the assessment of each safety review area, outlining:<br>- The main functions and objectives they cover<br>- How to produce structured and comprehensive assessment reports<br>- Identification of possible divergence and how to conclude on its significance to safety<br>- How to manage documents and RAIs | (cont.) | (cont.) |
| | Explain the fundamentals of graded approach and identify the most safety-significant areas | | |
| | Describe how the nuclear safety reviews are documented by area and FANR's consistent standard template format | | |
| | Apply appropriate communication to deliver findings and request further supporting analysis | | |
| | Perform/review a set of practical cases and conclude on adequacy and compliance | | |
| | Outline FANR procedures related to the review of simulation-based safety assessments | | |
| | Describe how the initial evaluation progress and RAIs are managed and monitored | | |
| | Explain FANR's procedures to ensure information to support licensing decisions related to nuclear safety and security is complete and fit for purpose in each stage of the process; FANR-Licensing Management Procedure:<br>- Reviewers' roles and responsibilities<br>- Action steps<br>- Outputs needed to execute the FANR licensing process | | |
| | Review the regulatory powers bestowed on FANR employees for nuclear licensing | | |
| | Indicate how the licensing process links to other regulatory processes (e.g. inspection) | | |
| | Give examples of activities that are exempt from regulation | | |

## II-4.CAPACITY BUILDING PROGRAMME ON THE REVIEW OF NUCLEAR INSTALLATION SITE SAFETY IN THE NUCLEAR REGULATORY COMMISSION, UNITED STATES OF AMERICA

The United States Nuclear Regulatory Commission (NRC) has administrative procedures that describe the basic requirements for its technical staff. The training provided incorporates implementation of the regulatory processes for nuclear facilities, equipment, processes and activities, as well as the criteria, techniques and mechanics of implementing nuclear reactor and power plant operations. The qualification process is intended to provide staff with sufficient information to regulate in accordance with NRC regulations, policies and procedures.

### II-4.1. Training for technical staff

#### *II-4.1.1. Regulatory requirements*

Regulatory requirements for design and operability of systems in a NPP are provided in the applicable sections of Title 10 of Code of Federal Regulations (CFR) part 50 (10 CFR 50) [II-22]. Appendix A to Ref. [II-22] (General Design Criteria for Nuclear Power Plants) provides specific elements that are applicable to NPPs. All technical staff need to be sufficiently familiar with these regulations. Reference [II-22] also includes important sections on 'Contents of Application; Technical Information', 'Technical Specifications', 'Environmental Qualification of Electric Equipment Important to Safety', 'Protection and Safety Systems', 'Requirements for Reduction of Risk From Anticipated Transients without Scram (ATWS) Events for Light-Water Cooled Nuclear Power Plants' and 'Risk Informed Categorization of Systems and Components'.

NUREG-0800, "Standard Review Plan for the Review of SAR for Nuclear Power Plants: LWR Edition" [II-23] provides guidance for review of the specific discipline (e.g. electrical, mechanical, civil, structural) portions of: (a) applications for nuclear reactor licences or permits, and (b) amendments to existing licences for each of the areas of review describing the review process and acceptance criteria. The SRP guidance may also be applied in the review of topical reports submitted to the NRC, especially reports requesting generic acceptance of systems or components that may be used in NPP safety systems. The SRP references industry standards, NRC Regulatory Guides endorsing standards, and Branch Technical Positions. The latter are prepared to resolve technical problems or questions of interpretation that arose during licensing reviews.

#### *II-4.1.2. General training*

In general, each employee responsible for technical reviews needs to meet the following qualification plan:

(a) General qualification requirements
(b) Position-specific qualification requirements, including for the following positions:
   — Reactor Technical Reviewer;
   — Reactor License Renewal Technical Auditor/Team Leader;
   — Reactor Environmental Officer;
   — Operating Reactor Licensing Project Manager;
   — Reactor Regulation Project Manager;
   — License Renewal Project Manager;
   — Operating Experience Engineer;
   — Reactor Oversight Process (ROP) Engineer;
   — Research and Test Reactor (RTR) Project Manager;
   — Rulemaking Project Manager;
   — Reactor Financial Reviewer;
   — Reliability and Risk Analyst;
   — Allegations Coordinator.
(c) Technical branches have supplemental requirements that include:
   — Branch specific courses and activities to expand knowledge beyond the basic qualification. The objective is for all technical reviewers to have a general understanding of all discipline related topics.
   — Branch specific courses and activities to support subject matter expertise in specific technical areas.
   — Branch specific periodic courses, training and activities for knowledge management, development, refresher training and optional training

#### *II-4.1.3. Training for inspectors of the United States Nuclear Regulatory Commission*

### II-4.2. **General training**

The NRC inspector training and qualification programme requires completion of a variety of activities which include:

(a) The legal basis for and the regulatory processes used to achieve the NRC's organizational structure and objectives;
(b) The basis for the authority of the NRC and its limitations;
(c) The procedures and processes established to achieve the regulatory objectives;
(d) Science and engineering fundamentals in the specific discipline of expertise, which include fundamental plant design and operation;
(e) Basic nuclear plant design and operations, which provide for protection of public health and safety;
(f) Specific reactor type or additional requirements for unique facilities which include:
  — Reactor full series boiling water reactor and pressurized water reactor;
  — Power plant engineering fundamentals.

### II-4.3. **Additional training on specific elements important for inspectors**

The inspectors are provided with a working knowledge of the contents of Appendix B of Ref. [II-22] (Quality Assurance Criteria for Nuclear Power Plants and Fuel Processing Plants), industry standards, and the associated licensee programs and documents that collectively establish the basis for the licensee's quality assurance programme.

There is also training on processes associated with review of risk-significant components to verify their initial design and subsequent modifications to determine their capability to perform their intended safety function(s) and identify performance deficiencies. This training provides the basis of failure modes of risk-significant components to perform the safety function and increased likelihood that the plant's design may not be able to mitigate the loss of the safety functions of those components.

## REFERENCES TO ANNEX II


[II-1] INTERNATIONAL ATOMIC ENERGY AGENCY, Establishing the Safety Infrastructure for a Nuclear Power Programme, IAEA Safety Standards Series No. SSG-16 (Rev. 1), IAEA, Vienna (2020).

[II-2] INTERNATIONAL ATOMIC ENERGY AGENCY, Governmental, Legal and Regulatory Framework for Safety, IAEA Safety Standards Series No. GSR Part 1 (Rev. 1), IAEA, Vienna (2016).

[II-3] INTERNATIONAL ATOMIC ENERGY AGENCY, Safety of Nuclear Power Plants: Design, IAEA Safety Standards Series No. SSR-2/1 (Rev. 1), IAEA, Vienna (2016).

[II-4] INTERNATIONAL ATOMIC ENERGY AGENCY, Safety of Nuclear Power Plants: Commissioning and Operation, IAEA Safety Standards Series No. SSR-2/2 (Rev. 1), IAEA, Vienna (2016).

[II-5] INTERNATIONAL ATOMIC ENERGY AGENCY, Safety Assessment for Facilities and Activities, IAEA Safety Standards Series No. GSR Part 4 (Rev. 1), IAEA, Vienna (2016).

[II-6] EUROPEAN COMMISSION, FOOD AND AGRICULTURE ORGANIZATION OF THE UNITED NATIONS, INTERNATIONAL ATOMIC ENERGY AGENCY, INTERNATIONAL LABOUR ORGANIZATION, OECD NUCLEAR ENERGY AGENCY, PAN AMERICAN HEALTH ORGANIZATION, UNITED NATIONS ENVIRONMENT PROGRAMME, WORLD HEALTH ORGANIZATION, Radiation Protection and Safety of Radiation Sources: International Basic Safety Standards, IAEA Safety Standards Series No. GSR Part 3, IAEA, Vienna (2014), https://doi.org/10.61092/iaea.u2pu-60vm

[II-7] INTERNATIONAL ATOMIC ENERGY AGENCY, Predisposal Management of Radioactive Waste, IAEA Safety Standards Series No. GSR Part 5, IAEA, Vienna (2009).

[II-8] INTERNATIONAL ATOMIC ENERGY AGENCY, Decommissioning of Facilities, IAEA Safety Standards Series No. GSR Part 6, IAEA, Vienna (2014).

[II-9] FOOD AND AGRICULTURE ORGANIZATION OF THE UNITED NATIONS, INTERNATIONAL ATOMIC ENERGY AGENCY, INTERNATIONAL CIVIL AVIATION ORGANIZATION, INTERNATIONAL LABOUR ORGANIZATION, INTERNATIONAL MARITIME ORGANIZATION, INTERPOL, OECD NUCLEAR ENERGY AGENCY, PAN AMERICAN HEALTH ORGANIZATION, PREPARATORY COMMISSION FOR THE COMPREHENSIVE NUCLEAR-TEST-BAN TREATY ORGANIZATION, UNITED NATIONS ENVIRONMENT PROGRAMME, UNITED NATIONS OFFICE FOR THE COORDINATION OF HUMANITARIAN AFFAIRS, WORLD HEALTH ORGANIZATION, WORLD METEOROLOGICAL ORGANIZATION, Preparedness and Response for a Nuclear or Radiological Emergency, IAEA Safety Standards Series No. GSR Part 7, IAEA, Vienna (2015), https://doi.org/10.61092/iaea.3dbe-055p

[II-10] INTERNATIONAL ATOMIC ENERGY AGENCY, Leadership and Management for Safety, IAEA Safety Standards Series No. GSR Part 2, IAEA, Vienna (2016), https://doi.org/10.61092/iaea.cq1k-j5z3

[II-11] NUCLEAR REGULATORY AUTHORITY, Regulation on Human Resources of Nuclear Regulatory Authority, NDK, Ankara (2019).

[II-12] NUCLEAR REGULATORY AUTHORITY, Training Catalogue, NDK, Ankara (2022).

[II-13] NUCLEAR REGULATORY AUTHORITY, Preparatory Training Programme for Technical Staff, NDK, Ankara (2015).

[II-14] NUCLEAR REGULATORY AUTHORITY, Training Programme for Candidate Inspectors, NDK, Ankara (2015).

[II-15] NUCLEAR REGULATORY AUTHORITY, Training Programme for Site Safety Assessment Group, NDK, Ankara (2018).

[II-16] INTERNATIONAL ATOMIC ENERGY AGENCY, Managing Regulatory Body Competence, Safety Reports Series No. 79, IAEA, Vienna (2014).

[II-17] EUROPEAN ATOMIC ENERGY COMMUNITY, FOOD AND AGRICULTURE ORGANIZATION OF THE UNITED NATIONS, INTERNATIONAL ATOMIC ENERGY AGENCY, INTERNATIONAL LABOUR ORGANIZATION, INTERNATIONAL MARITIME ORGANIZATION, OECD NUCLEAR ENERGY AGENCY, PAN AMERICAN HEALTH ORGANIZATION, UNITED NATIONS ENVIRONMENT PROGRAMME, WORLD HEALTH ORGANIZATION, Fundamental Safety Principles, IAEA Safety Standards Series No. SF-1, IAEA, Vienna (2006), https://doi.org/10.61092/iaea.hmxn-vw0a

[II-18] INTERNATIONAL ATOMIC ENERGY AGENCY, Objective and Essential Elements of a State's Nuclear Security Regime, IAEA Nuclear Security Series No. 20, IAEA, Vienna (2013), https://doi.org/10.61092/iaea.ajrj-ymul

[II-19] INTERNATIONAL ATOMIC ENERGY AGENCY, Leadership and Management for Safety, IAEA Safety Standards Series No. GSR Part 2, IAEA, Vienna (2016), https://doi.org/10.61092/iaea.cq1k-j5z3

[II-20] INTERNATIONAL ATOMIC ENERGY AGENCY, Application of the Management System for Facilities and Activities, IAEA Safety Standards Series No. GS-G-3.1, IAEA, Vienna (2006)

[II-21] WATZLAWICK, P., BEAVIN, J.H., JACKSON, D.D., Pragmatics of Human Communication: A Study of Interactional Patterns, Pathologies, and Paradoxes, W.W. Norton & Company, 1967

[II-22] UNITED STATES NUCLEAR REGULATORY COMMISSION, Domestic Licensing of Production and Utilization Facilities, 10 CFR Part 50, NRC, https://www.nrc.gov/reading-rm/doc-collections/cfr/part050/full-text.html.

[II-23] UNITED STATES NUCLEAR REGULATORY COMMISSION, Standard Review Plan for the Review of Safety Analysis Reports for Nuclear Power Plants: LWR Edition, NUGRE-0800, USNRC, Washington, D.C.

# ANNEX III EXAMPLE INSPECTION PROCEDURE

## III-1. EXAMPLE CONTENTS OF AN INSPECTION PROCEDURE

An example of the main contents of the inspection procedure is as follows:

1. Objective

2. Scope of application

3. Applicable references

4. General aspects
    - 4.1. Glossary (definitions)
    - 4.2. Responsibilities
    - 4.3. Qualification of inspector
    - 4.4. Criteria of inspection
    - 4.5. Inspection areas and items (facilities + technical capability of operators)

5. Implementation procedures
    - 5.1. Establishment of inspection plan
    - 5.2. Conducting of inspection (incl. entrance/exist meeting)
    - 5.3. Completion of inspection
    - 5.4. Follow-up management of inspection
    - 5.5. Inspection record keeping

6. Reference

7. Attachments: template of inspection plan, report, finding/recommendation

## III-2. EXAMPLE GUIDELINES FOR PREPARATION AND PERFORMANCE OF AN INSPECTION

### III-2.1. Establishment of an inspection plan

The inspector needs to establish an acceptable inspection plan that considers various factors, such as the field situation. The inspector has the discretion to establish the plan but is responsible for executing it efficiently as approved by the superior. The details of the plan may vary depending on the inspector's experience, capabilities, and the field conditions. General factors for establishing the plan include site conditions, target facilities of inspection, the scope of the inspection, and the safety category of the facilities.

### III-2.2. Notice of inspection plan

The inspector needs to notify the plant staff of the inspection plan in advance, using phone or other available communication methods. This gives them time to take necessary actions before the inspection. If the inspector cannot meet this commitment, the inspector needs to inform the plant staff about the delay and the reason for it in advance.

The notice needs to include the inspection's purpose, the expected duration, the participants, and any assistance needed from the plant staff.

### III-2.3. Review of pertinent documents

The inspector needs to thoroughly review the relevant documents and data before going into the field to perform inspection tasks efficiently. Generally, the relevant documents and data include the following, and may be adjusted based on the nature of the inspection:

- Safety analysis report and safety evaluation report of the plant;
- System description statement and design standard statement of the plant;
- Plant pre-operation inspection guidelines and prior references;
- Inspection records and administrative measures of precedent units;
- Other inspection records and administrative measures;
- Legislation, regulatory guidelines and technical standards for nuclear power;
- Field installation/test guidelines and applicable data.

### III-2.4. Inspector clothing and equipment

The inspector needs to dress appropriately for the tasks at hand, such as wearing a work uniform during field inspections or physical work, and plain clothes for meeting with management.

Before starting the inspection, the inspector needs to check that all equipment is functioning properly. The inspector also needs to carry the necessary forms and documents for recording information and handling administrative tasks related to the inspection.

### III-2.5. Changes to the inspection plan

The details of the inspection plan are those approved by the superior, and the inspector needs to promptly inform their superior of any changes, such as a change in the inspection target, for further instructions. However, if the changes are minor, the inspector is not required to report them before proceeding with the work, though they may do so at their own discretion.

If the inspector is unable to follow the established inspection plan due to changes in field conditions or other reasons, they may conduct an alternate inspection by gathering relevant information or reviewing other inspection records. The inspector needs to report the details of the alternate inspection to their superior once it is completed.

### III-2.6. Meeting prior to field inspection

The inspector needs to hold a meeting with the inspected entity before the field inspection to ensure efficient task performance. Meeting participants need to be limited to the Head and responsible staff of the entity, but no formal procedures are needed. At the meeting, the inspector needs to outline any needed assistance and listen to participants' input. The meeting details need to be documented.

### III-2.7. Performance of field inspection

The inspector needs to conduct field inspection tasks in a way that minimizes disruption to the inspected entity's operation. The typical steps of the field inspection are as follows, and may be adjusted at the inspector's discretion:

- Confirming the field situation;
- Reviewing the pertinent documents and records;
- Confirming the installations at the site;
- Interviewing the responsible staff at the site;
- Coordinating inspection team opinion for the inspection results;
- Recording the inspection results and defining actions to be taken;
- Developing the data for the field departure meeting.

### III-2.8. Issuance of inspection findings report and recommendations

The inspector may request correction of any problems identified during the field inspection, if the issues could impact plant safety or violate legislation or regulatory guidelines, the inspector may issue inspection findings report or recommendations.

#### *III-2.8.1. Inspection findings report*

The inspection findings report needs to include any violations of nuclear legislation, the SAR, the radiological EIA report, or applicable technical standards and specifications that, if unresolved, could compromise system reliability, functionality, safe operation, or radiation protection at the plant site.

#### *III-2.8.2. Inspection recommendations*

The inspection recommendations need to address work at the plant that, while unlikely to cause technical issues, are necessary for procedural or operational improvements.

### III-2.9. Meeting after field inspection

The inspector needs to hold a meeting after the field inspection to providing the entity being inspected with precise inspection results. At this meeting, the inspector needs to provide the entity with details of the problems observed, preliminary findings or recommendations derived from the inspection results. Also, the inspector needs to verify if further inspection is needed due to discrepancies with the opinions of the Head of the inspected department. All relevant plant staff need to attend the meeting to prevent any conflict in interpretation of the inspection results.

# ANNEX IV TRAIN-THE-TRAINER PROGRAMME

The necessity for capacity building continues throughout the lifetime of the organization. Training is necessary for new staff, for existing staff as they move to new positions and take on new responsibilities, as regulations change, or as the country acquires new obligations (e.g. new nuclear installations or new nuclear challenges). A cadre of personnel capable of providing the necessary training enhances the flexibility of the organization to provide timely, effective training to staff as needed. The trainers themselves therefore need to be trained in how to conduct effective training.

It is not sufficient for a trainer to have expertise in the topic on which training is to be provided, although that is a necessary attribute. In addition, the trainer is supposed to be able to assess the learning style of the trainee and provide the information in a manner in which it will be received, understood and learned effectively. The organization needs trainers that have flexible schedules and can travel, if necessary, to the locations of training needs. Owing to the breadth of technical topics in which training is necessary, it is difficult to find good trainers within the organization; this makes it important to develop an effective programme for training the trainers.

There are three steps an organization can take to develop capable trainers:

— Identify personnel with the interpersonal skills and necessary attributes to be an effective trainer;
— Select (or train) from among those personnel individuals with detailed knowledge of the necessary technical topics;
— Provide additional training to those personnel to enable them to be effective trainers.

This third step is the province of the train-the-trainer programme. The success of this step hinges on the first two steps being taken. Personnel selected to be part of the train-the-trainer programme need to have, or be able to acquire, the following:

— Outgoing personality, able to form a connection with a great variety of trainees, such as those who may be reticent to participate, be attending with reluctance or poor attitude, or who are concerned the material will be beyond their capabilities;
— An ability to 'read' people and form connections on their level;
— Skills in public speaking and presenting information in a clear manner, with respect to both written material and delivery;
— An ability to provide training in a variety of formats, such as lectures, role-playing, brainstorming, team meetings and mentoring;
— Willingness to invest in the success of the trainees;
— Knowledge of the attributes necessary to instil in the trainees;
— Enough understanding of the technical topics to appreciate the difficulties trainees may encounter;
— Skills in using feedback for continuous improvement of the programme.

Some organizations identify in-house personnel who can effectively provide training to the trainers. Alternatively, because these skill sets are independent of the technical topic involved, external train-the-trainer programmes are widely available. In either case, a programme for training the trainers generally includes the following:

— How to effectively interact with the trainees;
— How to provide training (preplanning, developing an agenda and the necessary training materials);
— Understanding the skills or knowledge to be acquired by the trainees;
— Feedback loop and follow-up.

## IV-1. TRAINING ON HOW TO BE AN EFFICIENT AND EFFECTIVE TRAINER

### IV-1.1. Attitude of trainer

A trainer's attitude towards providing training contributes significantly to the success of the training programme. If the trainer conveys an attitude of enthusiasm toward the subject material and confidence that the trainees can learn the material, the trainees are more likely to approach the training material believing that they can master it. An outgoing personality, a genuine interest in the trainees themselves, an unswerving commitment to the importance of the material to be conveyed and an attitude of overall professionalism are key indicators of success.

Train the trainer programmes are largely dedicated to conveying the importance of these attributes and providing specific training on how to handle interactions with the trainees.

### IV-1.2. Training materials

Effective training materials capture the trainee's interest by using a variety of formats, colour and graphics, and presented in a variety of ways, such as through videos, interactive discussions, practice sessions, student interaction, field trips, simulators, and e-learning modules. Since students have different learning styles, a variety of materials and techniques helps them absorb the material as effectively as possible. Additionally, providing a mix of materials and presentation techniques during the course of a day keeps the trainee attentive and engaged in the training. As the development of effective training materials is a specialized expertise, external assistance may be sought.

The older style text-based training materials are best provided as background for detailed concentrated study by the trainee outside of classroom training time.

### IV-1.3. Feedback and follow-up

One way to encourage participants to participate actively in the training programme is to solicit ongoing feedback. This helps assess the effectiveness of the training materials and trainers, and the overall design of the training programme. No training programme is expected to be static and needs to evolve as warranted.

When all trainees have access to computers, an online form is a quick and effective way to capture feedback, and results can quickly be summarized and analysed. Otherwise, paper-based feedback forms can be used. The programme designers consider whether the feedback will be anonymous, taking into account the pros and cons. Each organization needs to understand its own trainees in order to accomplish the goal of receiving honest feedback.

The feedback received is then used to modify the training programme so as to achieve continuous improvement.

## IV-2. IAEA TRAIN-THE-TRAINER COURSE FOR RADIATION PROTECTION OFFICERS

The IAEA train-the-trainer course for radiation protection officers [2] is aimed at developing communication skills as well as familiarizing participants with IAEA training material with a view to building a core of national trainers in radiation protection. The course syllabus includes presentation and communication skills, organization of training events and practical exercises. The course is designed to be interactive with an emphasis on including presentations by the participants. The course has been run around the world at both national and regional levels.

---

[2] https://www.iaea.org/services/training/train-the-trainers

The IAEA train-the-trainer courses help to equip Member States with competent radiation protection personnel, as mandated by the international safety standards. These courses cover a range of industrial and medical radiation practices, ensuring participants gain both theoretical knowledge and practical skills. Through modules on fundamental topics and supplementary sessions on specific applications, participants learn how to train others effectively, fostering the development of sustainable national safety infrastructure. The curriculum includes soft skills like communication and presentation, as well as technical sessions on radiation protection principles and the role of radiation protection officers.

## LIST OF ABBREVIATIONS

| | |
|---|---|
| BPTC | Basic Professional Training Course |
| CB Project | project on “Capacity building on site safety review and assessment in embarking countries” |
| CBAP | capacity building action plan |
| EESS | External Events Safety Section |
| EIA | environmental impact assessment |
| FANR | Federal Authority for Nuclear Regulation of the United Arab Emirates |
| HCB Programme | KINS Human Capacity Building Programme |
| INSS | International Nuclear Safety School |
| IMS | Integrated Management System |
| KINS | Korea Institute of Nuclear Safety |
| KSAs | knowledge, skills and attitudes |
| NDK | Nuclear Regulatory Authority of Türkiye |
| NEPIO | Nuclear Energy Programme Implementing Organization |
| NPP | nuclear power plant |
| NRC | United States Nuclear Regulatory Commission |
| OJP | on-the-job participation |
| OJT | on-the-job training |
| RAI | request for additional information |
| SAR | safety analysis report |
| SARCoN | Systematic Assessment of Regulatory Competence Needs |
| SEED | Site and External Events Design |
| SER | safety evaluation report |
| TSO | technical support organization |
| UAE | United Arab Emirates |

## CONTRIBUTORS TO DRAFTING AND REVIEW

| | |
|---|---|
| Al Senaani, H.S. | Federal Authority for Nuclear Regulation, United Arab Emirates |
| Aszodi, A. | Budapest University of Technology and Economics, Hungary |
| Contri, P. | International Atomic Energy Agency |
| Dubinsky, M. | Consultant, Israel |
| Gurcharan, M. | Nuclear Regulatory Commission, United States of America |
| Lee, H. | International Atomic Energy Agency |
| Lee, S. | Consultant, Republic of Korea |
| Sayin, B. | Nuclear Regulatory Authority, Türkiye |

**Consultants' meeting**

Vienna, Austria: 8–11 November 2022

**Technical meeting**

Vienna, Austria: 9–13 October 2023

IAEA
International Atomic Energy Agency

# CONTACT IAEA PUBLISHING

Feedback on IAEA publications may be given via the on-line form available at:
www.iaea.org/publications/feedback

This form may also be used to report safety issues or environmental queries concerning IAEA publications.

Alternatively, contact IAEA Publishing:

Publishing Section
International Atomic Energy Agency
Vienna International Centre, PO Box 100, 1400 Vienna, Austria
Telephone: +43 1 2600 22529 or 22530
Email: sales.publications@iaea.org
www.iaea.org/publications

Priced and unpriced IAEA publications may be ordered directly from the IAEA.

## ORDERING LOCALLY

Priced IAEA publications may be purchased from regional distributors and from major local booksellers.

25-00629E

FSC
www.fsc.org
MIX
Papier | Fördert
gute Waldnutzung
FSC® C083411

Zeitfracht Medien GmbH
Ferdinand-Jühlke-Straße 7
99095 Erfurt, Deutschland
produktsicherheit@kolibri360.de